Ferrocement Water Tanks
and Their Construction

S.B. WATT

PRACTICAL ACTION
Publishing

Practical Action Publishing Ltd
The Schumacher Centre
Bourton on Dunsmore, Rugby,
Warwickshire CV23 9QZ, UK
www.practicalactionpublishing.org

© Intermediate Technology Publications 1978.

First published 1978\Digitised 2013

ISBN 10: 0 90303 151 5
ISBN 13: 9780903031516
ISBN Library Ebook: 9781780442174
Book DOI: http://dx.doi.org/10.3362/9781780442174

Since 1974, Practical Action Publishing (formerly Intermediate Technology Publications and
ITDG Publishing) has published and disseminated books and information in support of
international development work throughout the world. Practical Action Publishing is a trading
name of Practical Action Publishing Ltd (Company Reg. No. 1159018), the wholly owned
publishing company of Practical Action. Practical Action Publishing trades only in support of
its parent charity objectives and any profits are covenanted back to Practical Action (Charity
Reg. No. 247257, Group VAT Registration No. 880 9924 76).

Contents

List of Figures

List of Tables

List of Photographs

Preface

This handbook has been prepared to describe to field workers and others how cylindrical water storage tanks of up to 150 cubic metres capacity can be built using wire-reinforced cement-mortar. The main advantages of this material over other tank construction materials, such as galvanised corrugated iron, are its cheapness and easy working using the minimum of expensive materials, equipment and skills. It is, in addition, very durable. Some of the tanks described in the manual have been in constant use for over 25 years with only a few instances of failure — due in the main to poor workmanship in construction. Galvanised iron tanks, even when properly made, have an expected life of only 5-10 years despite a much higher purchase cost.

The tanks may be built with a small capacity for domestic use, or larger for community water supplies, irrigation, stock watering and small scale industrial purposes. For domestic use the capacity is usually less than 10 cubic metres and may be as small as ½ cubic metre; the main requirement is that the tanks should be simple to construct by the users so that they can contribute their efforts towards the total cost of the tank. Extra tanks may be then added by the users at a later date if they see the need and have the resources. Self help tank building programmes can be equipped with soundly made formwork on which the tanks are built, which can be used for building many tanks. The extra cost of this equipment, which makes construction almost foolproof, may then be spread across all the tanks. Makeshift formwork is also used on isolated tanks but this increases the risks of poor construction and subsequent leakage. The larger tanks of up to 150 cubic metres capacity can also be built by self help but require greater organisation during their construction due to the greater quantity of reinforcing wire and mortar involved.

Storing water in tanks built on the surface has many advantages when compared with storage tanks excavated into the ground. Besides avoiding the need for laborious excavation which is almost impossible by hand in some hard dry soils, the tanks can be observed for leaks and easily repaired by trowelling a layer of mortar onto the inside of the empty tank. In addition, although the stored water is likely to become hotter in the sun, the risks of polluted material falling into the tanks are reduced. Water stored above ground can flow out under its own weight whereas it has to be pumped out of a ground tank.

The handbook has been divided into four parts:

Part 1 describes how the tanks are planned and designed. It gives the general sizes needed for different uses and an indication of the costs of these built in different parts of the world. A chapter is included on the ability of stored water to purify itself. The basic design is described to show how the reinforced mortar carries the loads when the tank is full of water.

Part 2 gives standard and recommended methods of constructing both small and large tanks using prefabricated formwork.

Part 3 describes various other construction methods that have been used in different parts of the world with the materials and equipment to hand; one of these is a successful commercial method. While some of these are not recommended because of cost and the complications they produce during building, they demonstrate the wide variety of ways that reinforced mortar has been successfully used.

Part 4 gives approximate calculations of the expected loads that the reinforced mortar must carry. It describes roof catchment water supplies in detail; sources of further information are also listed.

Hand trowelled, thin-walled cement-mortar water tanks are not difficult to build and may be used to store many different liquids and materials other than water. It is hoped that this handbook will encourage field workers to explore and utilise the wider possibilities of the materials.

S.B. Watt, C. Eng.

9

Chapter One
Introduction

Water tanks made from wire-reinforced cement-mortar are used widely in many parts of the world to store water for domestic, stock, irrigation and industrial purposes. They are built by hand trowelling a cement-rich mortar onto a mesh of wire reinforcement to form cylindrical tanks with thin walls which vary in thickness from 3 to 10cm depending on the size of the tank. The steel reinforcement usually consists of straight fencing wire wrapped during construction around a cylindrical formwork, or woven wire mesh tied to a supporting framework of weld mesh or heavier reinforcing rod.

Although this publication is called *Ferrocement Water Tanks and their Construction,* this is not strictly accurate. The tanks it describes should really be called wire-reinforced cement-mortar water tanks. The main difference is that in ferrocement there is a very dense mesh of woven or welded reinforcing wire that has to have a minimum value of wire volume for each unit volume of material. The quantities of straight wire reinforcement used in most of the examples collected for this publication fall far below this minimum value, although they provide ample strength for the purpose. Nevertheless, the material is closer in many respects to ferrocement than to ordinary reinforced concrete; this will be described in greater detail in Chapter 4. The wires distribute the loads through the mortar preventing them from concentrating in planes of weakness which would lead to the early failure of an unreinforced material. The straight wire reinforcement is chosen because it is both many times cheaper than the equivalent weight of woven wire and is easy to wrap around a small diameter cylindrical form.

This method of water tank construction is particularly suited for use in low income rural areas for the following reasons:

11

i) *Commonly available materials*
The basic raw materials of water, sand, cement and reinforcing wire are available in most areas and are already being used for many familiar purposes; in addition, except for the cement which must be kept dry, the materials are not easily damaged during transportation.

ii) *Simple skills needed*
The practical skills needed to use the materials are often known locally and untrained people can make satisfactory tanks after only a few days supervision.

iii) *Self help contribution*
The users of the tank can help in collecting sand for the mortar and in doing most of the heavy construction work; in this way they can contribute work instead of capital which they might find difficult to come by.

iv) *Simple equipment*
The construction techniques are simple and do not demand the use of expensive and sophisticated machinery nor a power supply; trained supervision can therefore be kept to a minimum. Leaks resulting from bad workmanship or damage can be simply repaired and maintenance after construction is negligible.

v) *Shared cost of formwork*
Where needed, the formwork used during construction can be made from cheap local materials such as timber or even adobe, or it can be prefabricated from more permanent materials to be carried from site to site and used to build many separate tanks.

Wire-reinforced cement-mortar is used for a wide variety of purposes but its particular advantage for water tanks is its ability to resist corrosion and its cheapness in comparison with other materials. Circular, corrugated galvanised iron tanks have been widely used in the past for water storage but these are expensive and corrode and burst within 5-10 years, even if they are carefully maintained. In contrast the life of a water tank made from reinforced cement-mortar is expected to exceed 50 years of continuous use.

The thin walls of the tank are able to deform under load and help prevent the early concentration of stresses that could cause failure of the tank. The dense wire reinforcement distributes the stresses through the mortar shell increasing its ability to carry the pulling and bending loads without cracking. This flexibility cannot be achieved in conventional steel-bar reinforced concrete because of the greater thickness involved.

Wire-reinforced cement-mortar tanks have been well proven in use over many years in the extremes of climatic conditions, and can be built confidently with capacities up to 150 cubic metres (about 33,000 gallons) although tanks of over 450 cubic metres capacity have also been successful.

Chapter Two

Water Storage and Costs of Construction

Size of tank needed

The size of water storage tank that is needed depends on the use to which the stored water is to be put. Small-scale water storage, even in the cheap tanks described in this publication, is still relatively expensive. This effectively limits the use to essential domestic water supplies, community water supply systems, irrigation water for high value crops, stock watering and industrial purposes. The tanks should not be built above ground on a tower unless specialist advice is available.

a. Individual domestic water supplies

In many areas of the world water that runs off the house roof is collected and stored for later use (see Fig.1). The roof has to be made from suitable durable materials such as clay tiles, galvanised iron or asbestos-cement sheets, etc. The roof acts as a catchment area to intercept the falling rain and has the great advantage that its cost is usually met during the construction of the house; this therefore subsidises the cost of the water supply.

The size of the storage tank needed will depend on the water demand of the residents and the amount and frequency of the rainfall. Chapter 16 describes how the tank size may be estimated. In wet tropical areas the small unreinforced jars of less than 0.5 cubic metres capacity described in Chapter 9 will make a major contribution to supplies, but in dry areas the water will be used up quickly and replenished less frequently. In these cases, a larger and more expensive tank of 10 cubic metres capacity or more will be required.

This type of water supply has the great advantage that it is unlikely to be contaminated and is within the control of

Fig. 1 Catching and using the rain water from a roof

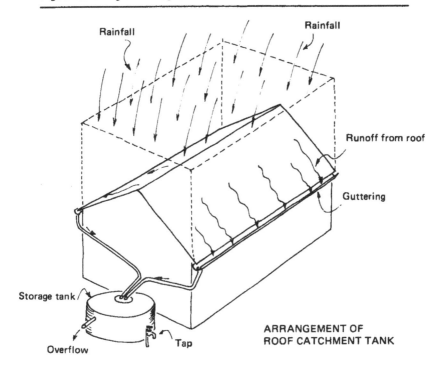

Rainfall

Rainfall

Runoff from roof

Guttering

Storage tank

Overflow

Tap

ARRANGEMENT OF
ROOF CATCHMENT TANK

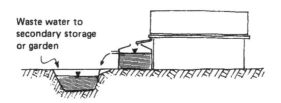

Waste water to
secondary storage
or garden

CROSS SECTION OF HOUSE AND TANK

the users to ration or use completely during the dry periods as they see fit.

b. Community water supplies

In hospitals, schools or community centres the advantages of the large roof and a large storage tank to catch the run-off can be shared between many people. Control of the rate that the water is taken from these communal tanks will be difficult and it is likely that it will be wastefully used unless water consumption is carefully planned. In areas that have large numbers of suitably roofed houses the individual tank to each household will probably be the cheapest and most acceptable choice.

The larger tanks can be used as header tanks in small water supply schemes; these keep the pressure in the delivery pipelines constant and will act as a 'buffer' during periods of high demand, refilling during periods of slack demand. The tanks are also used to store water collected from specially prepared ground catchments where the larger size of tank allows significant economies of scale (for a given height of tank the amount of material needed in construction per unit volume of storage decreases with increase in storage).

c. Irrigating high value crops

In parts of the world where farmers are growing high value crops for cash sale the expense of a water tank may often be justified. The water may be collected either from a catchment area or pumped by windmills from underground; it is then used sparingly as and when it is needed at the discretion of the farmer.

Storage tanks are particularly valuable for use with the wind powered water pumping systems where the greatly fluctuating wind speed gives an erratic output from the pump; the reservoir of stored water can then help to provide a continuous supply. An example is illustrated in Photo.1.

The use of costly stored water for subsistence crops is likely to be prohibitively expensive in most cases unless the water is used for essential 'survival irrigation' where without the water the crops would be totally lost in periods of drought.

16

Photo : Peter Fraenkel

Photo 1 Tank for stock watering
The streaks on the outside of the tank indicate that it has been leaking
— probably due to poor workmanship; this has not affected the
structural soundness of the tank.

17

d. Stock watering

Parts of the semi-arid areas of the world that are too dry for tillage agriculture can carry large numbers of grazing animals — provided that a reliable source of drinking water is available for them. This water is either pumped up out of the ground or is caught in specially prepared ground catchment areas. Reinforced cement-mortar tanks, either open or with sealed roofs to minimise evaporation losses, can provide the storage needed for a water supply; the outlet from the tank is usually automatically controlled to prevent wastage. In order to prevent overgrazing the tanks should be small, numerous and widely distributed, rather than large and concentrated at a few places.

e. Industrial use

There are many liquids used in industrial processes that need storage. Many of these can be kept in the tanks although a possible damaging reaction with the cement must be checked first. Some tank users have safely stored a wide range of liquids — ranging from wine to sewerage wastes.

Cost of construction

The final price of the tanks will vary according to local conditions but will depend on the following:—

Materials

The cost of the sand, cement and steel wire or mesh reinforcement.

Formwork

The cost of the formwork made either for one usage with temporary local materials, or of more permanent construction from steel sheeting and angle iron. The latter may be used many times.

Wages

The cost of wages for plasterers and labourers if the tank is not built totally by 'self help'

Supervision

The cost of supervision during construction.

Transportation

The cost of transporting the materials and supervisors.

Maintenance costs for these tanks after construction are usually negigible — they will give a trouble free life.

The cement and reinforcement must be carried to the site and in isolated areas this can add substantially to the total cost of construction. However, the materials are not susceptible to serious damage during transportation and, with the exception of the cement, do not need special protection. 'Straight' wire in coils will be very much cheaper to purchase and transport than the equivalent weight of woven or welded wire mesh.

The great advantage of these tanks in rural areas is the possibility of allowing the users to contribute their time and skills towards the costs of the tank by collecting all the sand necessary for the mortar and doing most of the trowelling work. This is in contrast to the galvanised iron tanks which are carried into the area of use from outside and erected by skilled workmen. The relative costs of tanks made with different materials are shown in Table 1 and a more detailed breakdown of costs for one of the larger tanks is given in Table 2. Both of these tables should be used with care as prices will depend not only on material and labour costs but also on the efficiency of construction.

Table 1 demonstrates the wide range of costs that may be expected when building these tanks in different areas and the economies of scale to be found in the larger tanks. The higher prices of the New Zealand factory-made tanks reflects the higher wage costs in that country; elsewhere, in the case of the Rhodesian self help tanks for example, self help construction reduces capital cost considerably even though the tanks take longer to build. Accurate local pricing of the tanks can only be carried out by comparing different building techniques and material costs in each area. The exceptionally low cost of the small Thailand water jar results from easy access to cement supplies and the absence of expensive wire mesh reinforcement in the design.

Table 2 gives the breakdown of actual and percentage costs for a large ($150m^3$) tank. It shows the relatively high cost of the skilled administration and transportation needed to construct a tank of this size. In this case the formwork was a small fraction of the total cost and would be reduced even further if its use could be spread over several tanks during a tank building programme.

19

Table 1: Typical Costs and Labour Requirements for Thin Walled Cement-Mortar Water Storage Tanks

Tank	Capacity (m³)	Materials Cost $(US)	Materials % Total Cost	Labour Cost $(US)	Labour Man days	Labour % Total Cost	Total Cost $(US)	Cost/m³ ($/m³)
1. Wire Reinforced Cement-Mortar Tanks								
Thailand water jar	0.5	0.5	100	—	1	—	0.50	1.0(1974)
New Zealand	4.5	NA	—	NA	2	—	93.0	23.3
Factory-made	13.5	NA	—	NA	3	—	210.0	15.6(1973?)
Tanks	22.5	NA	—	NA	5	—	280.0	12.4
Self-help domestic tanks, Rhodesia:								
Self help and supervision	9.0	62.5	66	30.0	20	34	92.5	10.3(1973)
Self help only*	9.0	62.5	100	—	18	—	62.5	6.9
Experimental tank built without shuttering, U.K.	6.3	55.0	100	—	15	—	55.0	8.7(1974)
Hans Guggenheim tank, Mali	10.0	NA	—	—	—	—	NA	NA
Rhodesian tank with roof	40.0	NA	—	—	—	—	NA	NA
U.S. tank built with								
with weld mesh	15.7	30.0	100	—	10	—	30.0	1.9(1965)
with woven mesh	139.0	98.0	100	—	15	—	98.0	0.7(1965)
Rhodesian tank with open top, Self help and supervision	150.0	199.0	32	246	34	40	612.0	4.1(1960)
	—	—	—	—	135	—	—	
2. Corrugated-iron Circular Galvanised Tanks								
Swaziland*	2.25	—	—	—	—	—	45.0	20.0
Swaziland*	4.50	—	—	—	—	—	65.6	14.6(1973)
Bulawayo*	9.0	—	—	—	—	—	100.0	11.1
Bulawayo*	9.0	—	—	—	—	—	112.0	12.4

NOTE:

Items marked with an asterisk* have been taken from *Catchment Tanks in Southern Africa – A Review* by A. Pacey, Africa Field Work and Technology Notes, Oxfam, U.K., 1974.

KEY: Man-days — the number of men working x number of days. NA — not available.

Table 2: Construction Costs and Bill of Quantities for 10m dia Tank with a Concrete Floor, 150m³ capacity, built 1960, Rhodesia

Item	Quantity	Rate $	Total Cost $SA	% Total Cost
Labour				
1. Trained foreman (17 days)	154 hrs.	1.0/hr.	154.0	25.2
2. 6 No. unskilled labourers (17 days)	1020 hrs.	0.65/hr.	66.3	10.8
(Partial self help) 1 No. skilled				
bricklayer (17 days)	170 hrs.	0.15/hr.	25.5	4.2
Total Payment for Labour			245.8	40.2
Walls and Floor				
3. Cement 100 bags	5000 Kg.	0.18/bag	80.0	13.1
4. Coarse sand	7m³	1.4 /m³	9.9	1.6
5. Fine sand (for walls)	5m³	0.6 /m³	3.3	0.5
6. Coarse aggregate (1½cm dia)	5.5m³	3.0 /m³	16.1	2.6
7. 39m x 1m pig netting	4 rolls	6.4 /rl	25.8	4.2
8. 370m rolls 4mm ø galv. wire	6 rolls	8.3 /rl	50.0	8.2
9. Soft iron wire	1 roll	8.2 /rl	8.2	0.3
10. Bitumastic primer	4.51	1.1 /drum	1.1	0.2
11. Bitumen jointing	45 Kg.m	–	3.0	0.5
12. Celatex for joint gaps	–	–	1.9	0.3
Total for Wall and Floor			199.3	32.5
Outlet Works				
13. 7.5cm sluice valve	1 No.	–	9.9	1.6
14. 7.5cm flange	1 No.	–	0.5	0.1
15. 7.5cm long bend	1 No.	–	2.3	0.4
Total for Outlet Works			12.7	2.1
Formwork etc.				
16. 2m high corrugated galvanised iron	–	–	47.5	7.8
17. Tools, etc.	–	–	19.2	3.1
Total for Formwork etc.			66.7	10.9
Transport and Plant Hire				
18. Transport	–	–	68.5	11.2
19. Water cart and concrete mixer	–	–	19.9	3.2
Total			88.4	14.4
Total Cost of Reservoir (Including Formwork)			$612.0 (1960)	100.0

NB: These costs have been taken from the article referred to in Chapter 17 in order to indicate the relative proportion of costs between labour, materials, formwork and transportation. The quantities of cement and reinforcing wire differ considerably from those calculated for the similar tank shown in Chapter 8 but they are included for consistency.

Sound, well made formwork makes the construction work simple and foolproof as the reinforcing wire can be wrapped directly onto it and tied in place. The formwork should be designed and made to be stripped down into simple pieces that can be easily transported from site to site. Well made forms will last many years and their cost for each tank will therefore be small.

For single isolated tanks the cost of prefabricated steel formwork is often prohibitive. Cheap, temporary forms can be made from local timber. Some tanks have even been built around adobe walls which have been erected and plastered smooth on the outside; the wire reinforcement and mortar is then built around this and the temporary adobe walls removed when the mortar has hardened.

Several methods are also described in Part 3 that avoid the use of the initially expensive formwork. These use either weld mesh or angle iron to stiffen and support the reinforcing wire during the application of the mortar. Weld mesh and woven wire mesh are relatively costly materials and it is doubtful if this method of construction would be cheaper than constructing a temporary makeshift form from local materials; solid formwork that will not move during construction is a highly recommended investment even when only a few tanks are planned.

Chapter Three

The Effect of Storage on Water Quality

The bacteriological and chemical quality of the water stored in the tank will depend initially on the quality of the water put into the tank. If the tank is not open to further contamination the stored water will in time purify itself of most of the harmful bacteria by natural processes. This process of self purification is used as an essential step in most modern water treatment works.

The water pumped from a protected underground source will invariably be free of harmful bacteria, although in some areas it may be undrinkable due to the presence of dissolved salts. Surface water supplies open to the atmosphere, birds, animals and human activity, are much more likely to be contaminated. These sources of supply should therefore be protected, and the risks of disease-causing bacteria entering the tank at any time must be kept to a minimum.

Water running off roofs into domestic tanks may contain wind blown dust, bird and animal droppings and, if the roof is flat and used as living space, human contamination. This is usually carried in the first flush of water running off the roof after a dry period; in some parts of the world it is common practice for the users to hold the downpipe away from the tank to allow the first polluted flush to run away to waste.

There have been very few measurements of the pollution load from roof catchments but the long history of their use suggests that the water can be safely used for drinking and domestic purposes with few health risks. Ground catchments that are open to animals and people are more likely to be contaminated and they should, if possible, be protected by a fence or wall. Material containing harmful bacteria on both roof and ground catchments is likely to be quickly dried and sterilised by the heat of the sun before it is washed off by rainfall, especially in the semi-arid parts of the world.

Diseases that are related to water and water use are major causes of sickness and death to people who do not have protected or purified water supplies, efficient sanitation and an understanding of basic hygiene. Any measures that reduce the risks of these diseases will therefore be of the greatest value to the water users.

The bacteria that cause some of the most serious or fatal illnesses, especially cholera, typhoid and diarrhea in children, cannot live for any appreciable time outside of the human intestine, and in such an unfavourable environment as stored water will die out almost completely within the first week leaving a few which can last up to a month or so. Diarrheal diseases need the presence of large numbers of bacteria in the water for the disease to be transmitted and even storage for a few days will substantially improve the quality of the water. Cholera and typhoid, on the other hand, can be transmitted by ingesting only a few bacteria and water storage of less than four weeks should not be considered to render the water harmless.

Most of the experiments that have been carried out in the past to determine how pathogens survive in water used heavily polluted water at a temperature of $10°-15°C$. If the stored water is relatively clean and the temperature is higher then it may be expected to be purified even more quickly. The exception to this will occur with grossly contaminated water that will probably quickly turn septic and start to smell; however, this will also be totally unpalatable to the users. In this instance there is a possibility that some bacteria will survive and even begin to breed.

Storing the water for several weeks will therefore destroy most of the main disease-causing pathogens and also give most sediment and organic matter time to settle out. The users will be at greater risk if they consume contaminated water within a week of the tank being filled; this can be overcome by boiling or otherwise sterilising the water, e.g. by chlorine tablets, before drinking, or by having two tanks which are used in turn. An alternative method of treating the water that has been suggested is to have a sand filter off-take at the bottom of the tank. This is described in greater detail in the next chapter.

The question of health and water quality is too large to be considered in greater detail in this publication; further infor- mation may be found in the references quoted in Chapter 16.

After the tank has been built the first volume of stored water will dissolve some of the compounds in the cement from the walls and become slightly alkaline. This is unlikely to affect the water to any great extent and will not occur on a second filling.

The water in the tank must of course be protected from any further possibility of contamination. As shown in Fig.2, a cover is usually necessary for this, and the tank must not be allowed to act as a breeding ground for malarial mosquitos. Household tanks especially must be protected and covered; overflow pipes should be covered with fine wire mesh to pre- vent the entry of insects and flies. The inflow to the tank should if possible be covered with a wire gauze screen; this will prevent rubbish from flowing into the tank with the water and will stop insects, flies and rodents from falling into the tank.

Water for animals and irrigation does not need to be so carefully protected, but steps should be taken to reduce the risks of contamination to the minimum.

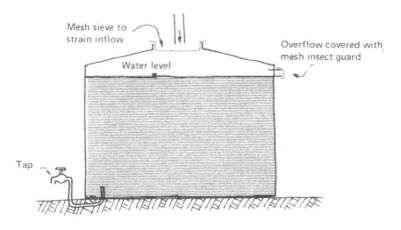

Mesh sieve to strain inflow

Overflow covered with mesh insect guard

Water level

Tap

Fig.2 Protecting the tank from further contamination

25

Chapter Four

Designing the Tank

The great advantage of wire-reinforced mortar over conventional reinforced concrete for water tank construction is its ability to resist shrinkage cracking during curing, its resistance to severe cracking under tensile load, and the need for only one set of forms for construction when the mortar is applied by hand to one side. Pouring a thin shell of concrete between two closely spaced shutters — the conventional method of reinforced concrete construction — is a highly skilled and difficult task.

Wire-reinforced cement-mortar has been used successfully for many years and indeed was used to make a rowing boat in the middle of the 19th Century, one of the first recorded instances of 'reinforced' concrete. It is more flexible in thin shells than the more conventional steel bar reinforced concrete and it is useful to describe in some detail its expected characteristics.

How does wire-reinforced mortar behave?

Unreinforced mortar and concrete are strong under compressive loads but very weak at resisting tensile or pulling loads; structures made from these materials that are subject to excessive tensile forces or bending will fracture suddenly without observable stretching and development of fine cracks (see Fig.3).

The weakness in tension and the brittle type of failure occur because however carefully the mortar is mixed and placed there will always be planes of weakness between the edges of discrete lumps that make up the mortar. These are exaggerated by shrinkage during curing and by imperfect bonding between each layer of mortar that is trowelled on. In

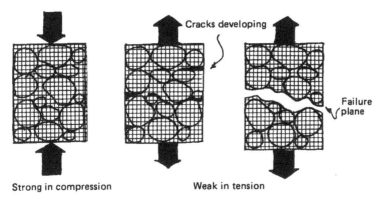

Strong in compression   Weak in tension

Fig. 3   The mortar under load

compression these planes of weakness are held together by the load, but under tensile loading they will open up beyond their elastic limit, coalesce with other cracks and rapidly cause the mortar to fail.

Conventional reinforced concrete is designed to overcome this characteristic by allowing the tensile loads to be taken completely on the reinforcing bars — the concrete in tension being assumed to have no strength. In reality, however, the reinforcing steel works to limit and control the tendency of the concrete to crack under tensile load according to the amount and distribution of the steel bars or wires and the degree of loading.

In reinforced cement-mortar under moderate tensile loads, such as those found in the small water tanks described in this handbook, the mortar may be assumed to contribute greatly to the tensile strength of the composite layer. This occurs because the wire mesh, distributed relatively densely through the mortar, will allow the load to be taken through-out the complete layer and will prevent the early concen-tration of critical stresses in planes of weakness. Any cracks that do appear under moderate loading will not be wide enough to allow water to reach the reinforcing wires and start corrosion.

The structural behaviour of a wire-reinforced mortar shell is difficult to calculate with any exactness especially if the wires, in the case of cylindrical tanks, are fixed mainly in

27

one plane around the tank. In addition, the mortar that is trowelled by hand onto the tank will inevitably be of varying thickness or strength. The calculations shown in Chapter 15, however, suggest that the smaller tanks are not highly stressed and there would seem to be a large factor of safety in most of the designs. This is demonstrated by the successful use of the tanks over many years.

The tanks described in this manual are in the shape of a cylinder with a flat floor and sometimes an integral roof. To achieve the greatest strength the tank should ideally be designed with the walls curved like a shell in both directions, horizontal and vertical (see Fig.4).

This will allow the water loading on the tank to be distributed over all of the tank structure and prevent critical breaking stresses from building up at an early stage at any one section. However, constructing tanks of this shape is very difficult and expensive; complex shaped formwork has to be used and the reinforcing mesh must be carefully cut, placed and tied down. Cylinders are very much more simple to make but the stress at the base of the tank where the walls join the floor is comparatively large and the joint must be made strong enough.

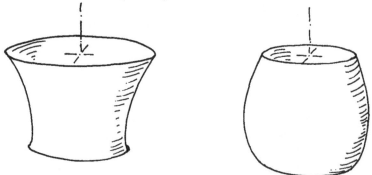

Fig. 4   Curved walls for greatest strength

The components of the tank

a. Foundation

The foundations of the tanks carry the weight of the tank and water down to the ground. The floor in the smaller tanks is usually continuous with the walls; the floor slab

28

carries the weight of the walls and the water directly on the foundation (see Fig.5).

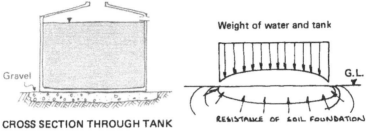

CROSS SECTION THROUGH TANK

Weight of water and tank

G.L.

RESISTANCE OF SOIL FOUNDATION

DIAGRAM OF FOUNDATION LOADING

Gravel

Fig. 5   Foundation of small tank — walls and base joined

The larger tanks usually have the floor built separately from the walls, and the floor slab therefore supports only the weight of water in the tank; separate foundations are needed for the walls (see Fig.6).

CROSS SECTION THROUGH TANK

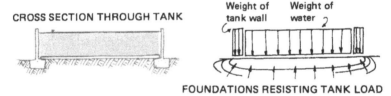

Weight of tank wall   Weight of water

FOUNDATIONS RESISTING TANK LOAD

Fig. 6   Foundation of large open tank — walls and floor separate

Preparing the foundations is one of the most important steps in tank construction and is considered in greater detail in later chapters.

b. Walls

The thin cylindrical walls, if they are free to move at the base when the tanks were full of water, would stretch under load to give only hoop tension forces within the walls (see Fig.7).

To prevent leakage, however, a flexible watertight seal of some sort would then be needed between the floor and the walls which will produce complications in design and construction. All of the tanks described in this publication have the walls built continuous with the floor or foundations. Although this produces some design difficulties it is an

29

almost universally adopted technique for the relatively small shallow tanks considered.

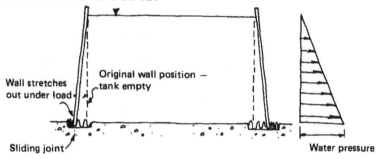

Fig. 7   *Walls free to move – exaggerated wall deflection*

The loaded tank will then deform roughly into the shape shown in Fig.8.

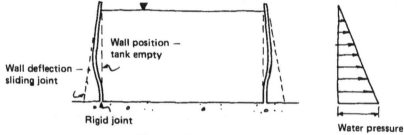

Fig. 8   *Walls joined with floor*

Analysis for the actual stresses of the water filled tank built in this way is very difficult as the design assumption of a homogeneous and uniformly elastic building material will not be achieved in practice. The analysis in Chapter 15 suggests that the vertical tensile stresses set up across a horizontal plane at the joint of the wall and the floor are nearly double the hoop tension stresses produced by the walls stretching outward under load. This indicates that the tanks should be specifically designed with vertical reinforcement both in the walls and between the walls and the floor to prevent cracking.

Most of the tanks described in this manual have hoop wire reinforcement with only a single layer of the more expensive woven mesh. The woven mesh, which is made continuous

30

between the walls and the floor or the foundation, will therefore provide the reinforcement that limits cracking at this joint. Experience with these tanks built under a variety of conditions suggests that severe cracks do not occur at the joint under normal loading.

The joint will not remain completely rigid, however. It is likely that very small cracks will open, and that the joint will bend outwards and throw a greater part of the loading onto the hoop wire reinforcement (see Fig.9).

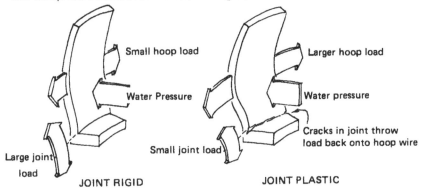

Fig. 9   How the loads are taken

These fine cracks are unlikely to be serious. If, however, they become wide enough to allow water to reach the reinforcement, because there is insufficient reinforcement or the wires are too widely spaced, corrosion and eventual failure will occur.

In most of the tanks a thick mortar coving is built up around the junction of the floor or foundation and the walls to strengthen the joint. The performance of this coving is difficult to predict but as there have been only a few recorded instances of tank failure, due mainly to poor workmanship, it would seem to prevent the cracking or limit it successfully. An alternative solution is to cover the junction of the wall and the floor with a bitumastic paint.

The woven chicken or pig mesh layer that is wrapped around the shuttering under the hoop wire in several of the designs will play an important part in preventing cracks occurring as a result of shrinkage and loading. Although it is

31

expensive it will contribute greatly to the soundness of the tank.

The tentative calculations given in Chapter 15 indicate that the stresses in a 2m high tank which is less than 10m in diameter with walls 5cm thick, will be small enough when full to be taken by a well made and carefully applied mortar with the minimum of reinforcement. In this case the reinforcing wire will behave mainly to strengthen the tank against external loads such as uneven settlement, earthquakes, temperature stresses, handling loads, etc. In larger tanks, however, the full amount of reinforcement is needed to take the greater stresses set up by the weight of the contained water.

c. Floor

The concrete floor of the tank may be built either before or after the construction of the walls. In small tanks (less than 5 metres in diameter) it is usually built first to give the walls a solid foundation and is made continuous with the walls. In the larger tanks (up to 10m diameter) it is also usually cast before the walls; a bitumen movement joint between the floor slab and the walls is provided to take expansion and settlement. Building the floor before the walls saves the work of lifting the concrete over the completed walls. In both cases, there should be wire reinforcement running from either the base slab in the small tanks or from the foundation of the larger tanks, up into the walls to provide resistance against cracking.

With larger tanks (over 10m in diameter) the concrete floor slab must be cast in sections with bitumen sealed expansion joints between the sections; this allows the floor to expand and contract without cracking. The floors of the smaller tanks described in this manual can, however, be built successfully as one continuous slab.

For cattle or irrigation water tanks the concrete floor may be replaced by a layer of worked clay and this will reduce the cost of the tank considerably. A suitable clay that can be 'puddled' or worked plastically is spread over the floor of the tank, moistened with water and rammed to compaction — a herd of goats has even been used for this. The clay floor must then be kept wet otherwise it will dry out and crack allowing the water put in the tank to leak out. The inner end of the outlet pipe for the tank must therefore protrude 10cm above the top of the clay to leave a small depth of water over the clay layer when the tank is emptied, this is good practice anyway. Alternatively, the concrete floor may be replaced by a sealed plastic sheet laid out to form a water proof membrane and tied into the base of the walls. It is protected both above and below by a layer of sand (see Fig.10).

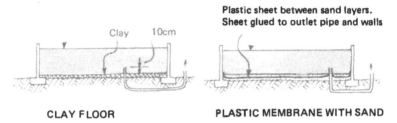

CLAY FLOOR   PLASTIC MEMBRANE WITH SAND

Fig.10 Alternative floors

d. Roof

A roof provides a good cover against evaporation, which in hot dry parts of the world can exceed 2m/year. It prevents the access of rubbish, insects and rodents and it also keeps the stored water cool. The tanks described in Chapters 10, 11 and 13 have integral shell roofs of wire reinforced mortar, between 3 and 5cm thick, cast continuously with the walls. Tanks of up to 5m diameter have been successfully roofed in this way and have stood up to long exposure in extremes of climate. The extra stresses set up by the weight of the roof are not great if the junction between the roof and the walls is curved; sharp angles concentrate stresses and initiate cracks. The greatest stresses set up in a roof of this sort are caused by its expansion in hot weather from the heat of the sun — these stresses can be over 20 times greater than the static loading stresses due to its own weight. For this reason the roof and walls should be painted white to reflect the heat of the sun or be protected with a covering of thatch (see Fig.11).

The roof may also be built from lighter weight materials, such as sheet aluminium or galvanised iron, fastened to a conventional structure erected over the tank (see Fig.12).

33

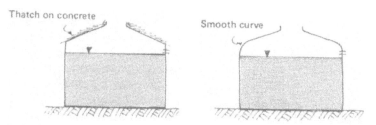

Thatch on concrete   Smooth curve

CURVE BETWEEN ROOF AND WALL   ROOF THATCHED OR PAINTED WHITE

Fig.11 Roofs for tanks less than 5m diameter

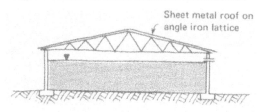

Sheet metal roof on
angle iron lattice

Fig.12 Roof for larger tank

e. Tank fittings

Most tanks will have pipe fittings of some sort to take the water in to and out of the tanks. The outlet pipe for the smaller tank is sometimes taken through the walls; the outflow pipe through the top of the wall; and the water enters through the top opening. With the larger tanks, however, and the larger volumes of flow that their use dictates, the inlet and outlet pipes are taken through the floor. This reduces the risk of cracking and leakage around the pipes which would obtain if the pipes were cut into the walls. The overflow may still be taken through the top of the wall. In all cases where the outlet pipe passes through the floor a wire screen of some sort is recommended, to prevent the outlet pipe being accidentally blocked by debris.

The outlet pipe mouth is usually placed 10cm or so above the level of the floor to prevent sediment that has settled from being carried out of the tank. The inlet pipe may also be taken in through the floor in which case the optional clay floor of a larger tank must be protected against scour.

The hole for the overflow pipe to the tanks is cut out of the mortar while while it is still 'green', i.e. before it has set

34

properly. The pipe, which must be large enough to carry away the anticipated overflowing water, is set into this hole; plastic pipe is not suitable because it is not rigid enough. With covered tanks used for domestic purposes the pipe must be covered with a layer of fine mesh to prevent malarial mosquitos or insects from entering and breeding in the tank.

On the larger tanks used for cattle watering the outlet is often provided with an adjustable pipe section fitted inside the tank; if, therefore, the water trough is damaged by the animals or if the ball valve in the trough sticks open, only a small amount of water will be lost (see Fig.13).

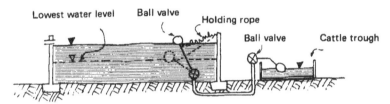

Fig.13 Ball valves for cattle watering

The outlet to the tank may be fitted with a sand filter that screens and sieves the water before it is used, removing most organic matter and bacteria. It will behave like a slow sand filter and must be kept permanently wet if it is to work correctly (see Fig.14).

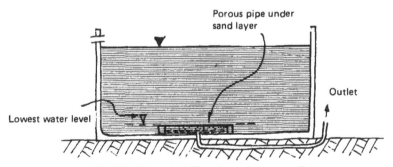

Fig.14 Slow sand filter in tank over outlet

The sand can be cleaned or replaced when the tank is empty.

35

Chapter Five

Construction Materials and Equipment

a. Reinforcing mesh

There are many different types of steel reinforcing mesh that can be used to construct reinforced cement-mortar shells. These generally consist of thin wires, either woven or welded into a mesh, but the main requirement is that it must be easily handled and, if necessary, flexible enough to be bent around sharp corners. The wires are tied and held firmly in place while the mortar is being trowelled on and should finish with an even distribution through the complete thickness of the shell.

Woven and welded mesh are expensive to buy, because of the work needed to make them, and are expensive to transport because of their bulk. They can cost over ten times as much as an equivalent weight of 'straight' coiled wire.. For this reason straight wire is most often used to construct the cylindrical water tanks described in this manual — it can be wrapped around the tank without difficulty. Reinforcement is still needed in a vertical direction, however, and this usually takes the form of a single layer of woven mesh wrapped onto the formwork before the hoop reinforcement is wound on. For the larger tanks the calculations in Chapter 15 indicate that extra vertical reinforcement is as important as the hoop wire, but for the smaller tanks (less than 5m diameter) the single layer of woven mesh is adequate.

The wire may be galvanised to prevent rusting during storage; under no circumstances should aluminium painted wire be used as the aluminium may react with the cement to give a very poor bond between the wire and the mortar.

b. Cement, sand and water

The cement that is used to make the mortar should be an

ordinary Portland cement (to BS12 or similar specification). Lower strength cementitious materials have been used in the past with some success but these generally cannot be recommended. The tank described in Chapter 14 has lime mixed with the Portland cement in the ratio of 1 bag of lime to 5 bags cement to improve the workability of the mortar and to reduce shrinkage cracks; it is not known if this reduces appreciably the strength of the mortar. The cement should always be kept in a dry store until it is to be used.

The main requirement for the sand is that it should be free from organic or chemical impurities that would weaken the mortar. Most clean sands are suitable but if their quality is in doubt they should be washed with clean water. They should be protected at the mixing site against further contamination.

A silica sand is probably the best although sands consisting of other hard minerals can be satisfactory. Experience suggests that a moderately coarse sand, although it makes the mortar more difficult to work, will resist shrinkage cracking better than a fine or dirty sand; if the sand has a high silt content then the mortar will be weak. The grading of the sand particles should make for an easily worked mortar and there should be a reasonable proportion of all grain sizes without an excess of fine or coarse particles.

The water must be fresh and free from chemicals in solution or suspended silt and organic matter. Clean water is essential for a strong durable mortar; salt water should never be used.

Keeping good control over the quality of the materials used for constructing the water tanks is the first step in developing sound construction practice.

c. The mortar mix

Making up a strong and satisfactory mortar mix from cement, sand and water is one of the most important stages in building the tanks. The mortar must be prepared with the correct proportions of these materials. It must be well mixed and workable enough to be trowelled by hand onto the formwork between the reinforcing mesh to form a dense, compacted layer, and it must be properly cured in order for it to achieve its full potential strength and durability. It is useful to understand the various factors that control the final quality of the mortar in order to understand the need for sound practice during construction.

Ratio of cement and sand (cement:sand by volume)

Increasing the proportion of cement in the mortar will increase its final strength and make it more workable, but it will lead to a greater risk of wide shrinkage cracks which might cancel out this increase of strength. It will also increase the cost.

The tank designs described in this handbook use a mix with a cement to sand ratio of between 1:2 (i.e. 1 part *by volume* cement:2 parts *by volume* of sand) for the small unreinforced water jar and 1:4 for some of the larger reinforced tanks. Some authorities recommend that for tanks of less than 40 cubic metres capacity, the proportion of sand should never exceed 3 times that of the cement. It is possible that this ratio is designed to give a large factor of safety, but the experience gained in field construction over the last 25 years suggests that the cement:sand ratio can be increased to 1:4 when using suitable sands without risk of failure. Making the cement:sand ratio 1:3 will not give a substantial increase in the cost of a small tank, however, and this ratio is therefore generally recommended.

Measuring boxes or buckets should be part of the construction equipment and should always be used to achieve consistency in mix proportions. Measuring the materials on a shovel does not give reliable results.

Ratio of water and cement (water:cement by weight)

A dry mortar mix will be stronger than a wet mix made with the same proportions of cement and sand provided it is fully compacted. Dry, stiff mixes are, however, difficult to work onto the formwork to achieve full compaction, are likely to contain air voids and be imperfectly bonded to the reinforcing wire. Wet creamy mixes are very easy to trowel by hand, but the cured mortar will be more permeable to water and have a lower strength and durability.

A compromise between aiming for either a strong or an easily worked mix must therefore be made. Experience has shown that with reasonable sands in a 1:3 (cement:sand) mix, a water:cement ratio of 0.5:1 (½ part *by weight* to 1 part *by weight* of cement) will be satisfactory. If, in order to make the mix workable enough, extra water has to be added to give a water:cement ratio of 0.6:1 then a better graded sand or a greater proportion of cement should be used.

Under most conditions the workability is usually controlled by eye during mixing and, if the water:cement ratio is not to exceed 0.5:1 (water:cement *by weight*), the mixing must be carefully supervised. The amount of water to be added to the dry mix of cement and sand is affected by the presence of moisture in the sand, which can vary considerably between the top and bottom layers of the sand stockpile; less water should be added the wetter the sand. One or two trial mixes using a sample of completely dry sand and a measured water:cement ratio will allow the people engaged in mixing to see and learn the feel of a satisfactory mix.

d. Formwork

Formwork is needed to support the walls while the mortar that has been trowelled on hardens and sets. Well made formwork is expensive but will, with care, last for many years and its initial cost may then be spread over many tanks. Several tank designs are included in Part 3 that use makeshift formwork or use weld mesh frames to support the wire mesh and trowelled mortar. However, experience shows that sound formwork makes construction work almost foolproof, and is therefore recommended for any tank programmes building more than a few tanks.

The reinforcing wire and mesh can be wound simply and quickly around the cylindrical forms and the mortar may then be trowelled over the reinforcement. The main requirement of the formwork is that it should be rigid enough to hold the weight of the mortar as it is being applied and cured without deflecting. If it does move during the setting period, the mortar is likely to crack and be considerably weakened. It will also help to limit the moisture loss from one face as the mortar hardens.

The formwork may be made from many different materials. The circular corrugated galvanised iron sheets shown in Chapters 7 and 8 have been used with great success. The corrugations provide a mark for the tank builders to help them wind the reinforcing wire on at the correct spacing although the wires will tend to bundle together in the corrugations. The main advantage of the corrugated sheets — besides durability, cheapness and lightness in transportation — is that they allow an accurate measure of the final wall thickness, because the corrugations on both the inside and outside faces of the tank must be filled with mortar and trowelled smooth. This is of great importance in self help construction as it reduces the need for skilled supervision and the risks of thin weak spots in the tanks walls.

Temporary formwork may also be made from suitable local materials such as rough sawn timber planks, securely braced, and some of these are described in Part 3. A temporary circular walling of adobe (mud and straw bricks) has also been used with success. This is built up to the required shape and plastered smooth on the outside with a weak mortar mix. When it has set, it is treated with a suitable separating oil and the wire and mortar built onto its outside. After the mortar has hardened, the adobe is broken up from the inside and removed from the tank. The example described in Chapter 12 uses a reinforced mortar lining inside a permanent, traditional adobe grain bin which both supports the lining and helps to contribute to the strength of the tank.

e. Tools

A list of necessary tools is outlined in Table 3 and should be collected to make a standard kit for tank construction programmes. It is possible to use local tools and equipment if they are available but it is always safer to make sure that all necessary tools are on hand by bringing in the standard kit.

The mortar may be mixed either by hand or with a concrete mixer. Hand mixing is laborious and if incomplete will give a lumpy mortar that is difficult to trowel onto the tank. Hand mixing can manage drier mixes than the concrete mixer, which rolls a dry mortar in its drum without proper mixing. Treading the mortar under foot on the mixing slab is one of the best ways of working a dry mix. For the tanks described in this manual the wetter mixes from a powered concrete mixer will be satisfactory.

Steel (plasterers) floats are much easier to use than those made of wood. Overworking the wet mortar, especially with a steel float, can cause the mortar layer to slump off; the mortar should be laid on quickly and carefully. The surface of each mortar layer must finally be roughed or brushed when it has hardened a little to provide a key for the next layer.

Table 3 List of tools and equipment

Number	Item
4	Plasterer's steel hand floats
4	Hand hawks 30 x 30cm with 4cm diameter handle
4	Trowelling boards, 75 x 75cm
2	Wire brushes for cleaning shuttering
2	Oiling brushes
1	Brush or scratching tool for mortar
1	Hacksaw and spare blades
1	Woodsaw
1 set	Spanners
1	Crowbar, 1 metre long
1	Wire snips for mesh
1	Bolt cutters for hoop wire
2	Cold chisels for cutting 'green' mortar
1	2 kg hammer
1	5 kg sledgehammer
1	Gauging box for sand and cement 50 x 50 x 40cm or container to hold 100 litres sand
1	Sieve 5mm maximum openings for sand
4	Shovels, flat ended for mixing
1	Spirit level
1	50m cloth or plastic tape measure
1	String line
1	Axe for trimming timber
2	Pickaxes for excavation
2	Mattocks for ground levelling
1	Wire tensioning tool
1 set	Formwork sections
1	Wheelbarrow for carrying mortar
	Water containers
	Plastic sheeting or hessian for curing the mortar

Chapter Six

Summary of Methods

This part of the handbook contains a detailed description of standard tank construction practice that has been used successfully for over 25 years in different parts of the world. Designs for a small 10m^3 capacity domestic tank and a large 150m^3 open tank are given in Chapters 7 and 8 respectively, and these are recommended for self help construction in rural areas. The designs have evolved from practical experience and the dimensions and quantities of reinforcement should be followed as closely as possible.

A summary of the basic steps in building these tanks is given below.

a. Clearing the site and preparing the foundations

The site chosen for the tank should be cleared of vegetation, loose surface soil and any large rocks that could pierce the floor of the tank. For the smaller tanks it is often only necessary to clear the site and put down a 15cm layer of sand and gravel ready for the tank after all the pipe fittings have been laid. The larger tanks, however, need a separate ring foundation to support the walls which is prepared by digging a trench under the line of the wall and backfilling it with concrete.

b. Collecting the materials

The sand, gravel and water should be collected before construction starts. If this is done by the users they will be able to contribute a large proportion of the total cost of the tank (and is one reason why these tanks are chosen for self help construction).

The aggregate can be stored in piles alongside the mortar mixing slab which is built by trowelling a layer of concrete onto a gravel layer 2m x 2m square. The slab is finished with a shallow wall or lip to prevent the accidental spillage of cement slurry during mixing (see Fig.15).

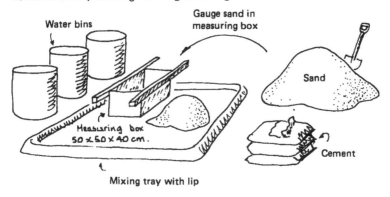

Fig.15 Mixing the mortar by hand

For isolated, small domestic tanks, it may be cheaper to carry a mixing board as part of the basic kit. Mixing the mortar on the ground surface will contaminate and weaken the mortar, and should always be avoided.

The aggregate piles must be kept clean of soil and other material, and are often covered with a plastic sheet or sacking to protect the sand from contamination by animals or wind blown dust. Make sure that enough sand is collected for each tank before construction begins.

The cement and reinforcement should be collected and stored under cover. The shuttering and any other equipment and tools are also collected and checked ready for use. Any pipework is installed before the concrete is laid.

The layout of the site for the larger tanks should be arranged to reduce unnecessary work. The cement, sand and water should therefore be stored near the mixing slab or concrete mixer on one side of the tank site. An approximate idea of the quantities of material needed for each tank is given in the chapters below; it is advisable to have extra material on hand to allow for wastage.

c. Casting the foundations
The foundations for the small tanks built on the site are made by casting a 7.5cm thick slab of concrete 2.8m in

43

diameter onto the layer of sand and gravel. The concrete is prepared from a mix of 1:2:4 (cement:sand:gravel by volume) and is given a week to harden.

The ring trench of the larger tanks is filled with concrete after the floor slab has been cast, and this is also allowed to harden. If foundation beams are needed under joints in the floor slab (for tanks of diameter larger than 10m) these are cast at the same time.

d. Erecting the formwork

The formwork is examined for damage, cleaned and oiled with a suitable release agent to prevent the concrete from sticking. Old motor oil can be used for this purpose. The formwork is then erected and bolted together in the way described in the chapters below. The recommended forms are built from rolled corrugated iron sheets.

e. Fixing the reinforcing wire and mesh

The woven mesh is wound around the outside face of the formwork and tied firmly into place. For the small tanks it is tucked under the formwork to join into the floor slab cast after the walls. In the larger tanks it is allowed to hang in the wall foundation trench which is then filled with concrete; it may with advantage be substituted in the larger tanks with 6mm diameter reinforcing rod at 10cm spacings.

The hoop wire is then carefully wound around the tank and spaced into the corrugations. Any joins in the wire should be overlapped by at least ½ metre and tied securely with soft iron wire; they should not be tied by wrapping the soft iron wire along the join as this will prevent the mortar from making a proper bond. If the hoop wires are slack they can be tightened by kinking them with a special tool (see Fig.16).

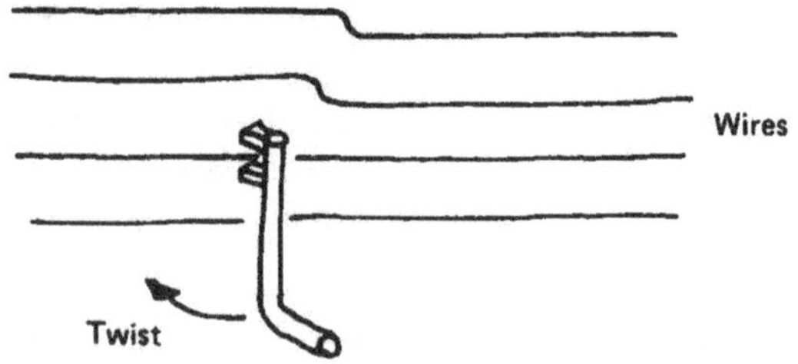

Fig.16 Tool for tightening wires

f. Mixing the mortar

The sand and cement are gauged either onto the mixing slab or into the concrete mixer in the ratio 1:3 (cement:sand by volume). It is difficult to judge these volumes accurately by shovelling and a gauging box should be made up to measure the sand. The box is made to the size 50 x 50 x 40cm; 1 box full of sand mixed with a 50 kg bag of cement will give a mortar mix 1:3 (cement:sand).

The dry cement and sand on the mixing slab are turned by hand (if a concrete mixer is not available) several times into piles from one side of the slab to the other until they are completely mixed together. A hole is then made into the centre of the pile, water is added and turned into the mix until the desired consistency is achieved. Add only a small amount of water at a time as it is difficult to make a wet mix drier by adding cement and sand. The workability of a dry mix can be increased by treading it between the feet to break down any drier lumps.

Conventional powered concrete mixers are often condemned for true ferrocement construction because they can only manage wet mixes, but for the tanks described here machine mixing will be quite satisfactory.

g. Trowelling the mortar onto the tank walls

After mixing, the mortar must be applied quickly to the tank; if it is more than ½ an hour or so old, it should be either used to make up the floor slab or thrown away as the cement quickly begins to become unworkable without adding an excessive amount of water. In hot climates the mortar pile should be covered with wet sacking or black plastic to prevent rapid drying out.

The mortar is applied by hand to the walls of the tank with a plasterers' steel float in layers not greater than 1cm or so thick; thicker layers will tend to slump off. The mortar which has already been prepared at the mixing slab is carried to the side of the tank and emptied onto a square board (75cm x 75cm) which may be made of plywood or planks. This board prevents dirt from becoming mixed up with the mortar, and catches any mortar that falls off the wall during trowelling. The mortar is scooped off this board using the plasterers' float onto a builder's 'hawk' (a square board 30 x 30cm with a 4cm diameter handle attached to the middle of one side) and transferred back onto the face side of the float (see Fig.17).

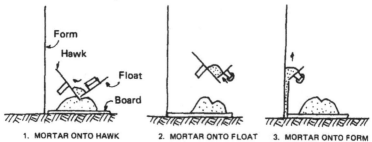

1. MORTAR ONTO HAWK   2. MORTAR ONTO FLOAT   3. MORTAR ONTO FORM

Fig.17 Trowelling the mortar onto the formwork

The mortar is trowelled onto the formwork from the base of the tank upwards to fill the corrugations and just cover the reinforcing wire. The walls are built up in this way in vertical sections around the tank. The square board is moved for each section.

When this first layer of mortar has hardened sufficiently, the surface is brushed rough or scratched ready for the second layer. This 1cm thick second layer, which provides the outside surface to the tank, is finished with a smooth surface. It must be well bonded to the first layer, which should be 'green', i.e. not hardened completely.

The next day, the formwork is stripped out from inside the tank, and a layer of mortar is trowelled on to fill up the corrugations and cover up any exposed reinforcing wires. Corrugated iron formwork has the great advantage that the correct thickness of the tank walls can be easily maintained even by unskilled workers.

If the tank has to be left overnight with a layer incomplete, the edge of the mortar should be cut off square. Next day, the joint should be brushed with a wire brush and be coated with a cement slurry to give a strong bond before the applic- ation of fresh mortar. It is advisable to complete any one layer in the same day, or if interruption is anticipated, to finish the layer in a complete band around the tank. Joins will then only occur in a horizontal line around the tank;

vertical joins which are more likely to open up should be avoided.

Applying the mortar onto the tank may appear difficult at first, but most people can learn the knack of doing it within a few hours. The secret is to apply the well mixed mortar quickly and firmly with the float, and work it smoothly over the wall. The surface finish is not of the greatest importance on these tanks; if they are rough they can be filled in at a later date. The layers must be of uniform thickness throughout, however, with no gaps or weak spots.

h. Curing the tank

After the mortar has been applied to the tank walls it should be covered up with black plastic or wet sacking. If the setting mortar is exposed to sunlight or wind it will quickly lose its water and the final strength and durability will be considerably reduced. Loss of moisture in this way exaggerates the shrinkage cracks that will form in the mortar and can even lead to tank failure.

In very hot climates the tank must be covered up between the application of each layer. In milder conditions or under cover the tank is left open until the final layer has been applied and it is then covered for a week or more to cure. The mortar will take at least one month to gain anything like its final strength, and for the first few days it will be 'green' enough to hack out by hand for any pipe fittings.

Curing is absolutely essential for sound, strong tanks, and is one of the most important construction steps. It is also one of the most difficult things to ensure in the field.

i. The floor and roof

The floors of the tanks are poured before the walls have been built. The roof of the smaller tank is built several days after the walls have been built to allow them time to strengthen.

j. Filling the tank with water

Cement mortar shrinks as it dries out in an empty tank especially in hot dry climates. If the tank is then filled rapidly with water the mortar does not have time to expand again as it reabsorbs moisture slowly, and the wire reinforcing will not contribute to withstanding the stresses in the walls. In this case there is a great risk of severe cracking or even failure. An empty tank, especially one that has just been built, should always be filled slowly, and it should be left for a week or so with a shallow depth of water at the bottom before filling.

Chapter Seven

Small Tanks for Domestic Use: 10m³ Capacity

These tanks have been used for many years in parts of Africa and have been designed to be as simple as possible to build in self help programmes. The users, who are at first unskilled in this sort of construction, can contribute their time and efforts in collecting sand and water, digging the foundations and preparing the mortar under the general guidance of a trained builder. With experience they quickly learn how to make the tanks without further guidance.

A trained builder with five helpers takes about three days to construct the tank. The users often contribute some money towards the cost of the tank, which helps to cover the builders' wages, the cement, reinforcement and the hire of the formwork.

Design

The tanks have been designed for construction by relatively unskilled workers. They have a diameter of 2.5 metres, a height of 2 metres, giving a capacity of 10 cubic metres. The final wall thickness will be about 4cm. The tanks are built on site and should not be moved.

Formwork

The 2m high formwork is made from 16 sheets of standard galvanised roofing iron, 0.6mm thick with 7.5cm corrugations, rolled into a cylinder with a radius of 1.25m.

Steel angle iron (40 x 40 x 5mm) is bolted vertically on the inside face at the ends of each set of four sheets — this allows the sheets to be bolted together to form a circle. Between the ends of each section is placed a wedge which is pulled out to allow the formwork to be dismantled (see Fig.19).

49

Fig.18 Standard formwork to make 10m³ tank

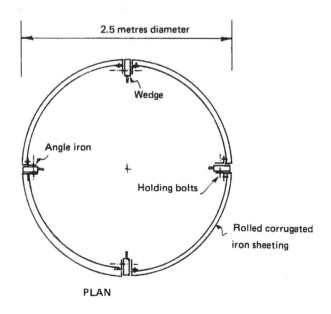

2.5 metres diameter

Wedge

Angle iron

Holding bolts

Rolled corrugated
iron sheeting

PLAN

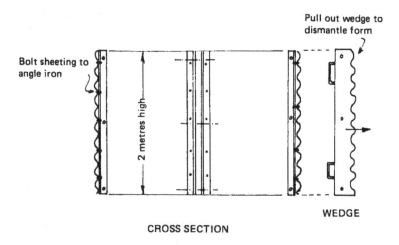

Pull out wedge to
dismantle form

Bolt sheeting to
angle iron

2 metres high

WEDGE

CROSS SECTION

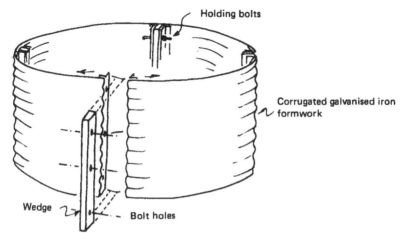

Fig.19 Assembling the formwork

Construction

A circular area 2.8m in diameter is cleared at the required site for the tank and excavated down through the loose topsoil. A 10cm layer of sand and gravel is laid evenly over the excavation and a 7.5cm layer of concrete laid on top of this; the concrete mix of 1:2:4 (cement:sand:gravel by volume) will form the foundation slab under the tank.

Into this concrete foundation is cast a 1m length of 20mm bore steel water pipe with a tap on the outside end. The pipe is curved so that it projects 10cm above the floor of the tank; a piece of wire is threaded through the pipe to act as a pull through after the tank has been built (see Fig.20).

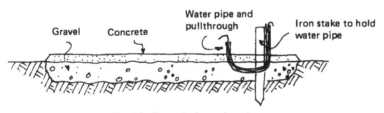

Fig.20 Foundation of tank

When this concrete floor slab has hardened the formwork for the tank is erected. The bolts passing through the angle

51

iron and wedges are tightened to provide a rigid cylindrical form. This is cleaned free from cement and dirt, oiled and the wire netting wrapped around it to a single thickness and tucked under the forms. The netting has a 50mm mesh, and is made from 1.0mm wire (see Fig.21).

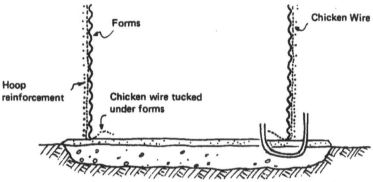

Fig.21 *Erect formwork and wind on reinforcing wire*

To form the hoop reinforcements, the 'straight' galvanised iron wire, 2.5mm diameter, is wound tightly around the tank from the base at the following spacings:—

2 wires in each corrugation for the first eight

1 wire in each corrugation to the top

2 wires on top corrugation.

About 200m of 2.5mm diameter wire will be needed, weight 8 kg. The netting provides vertical reinforcement to the tank and also holds the hoop wire out of the corrugations.

The outside is then plastered with a layer of mortar made from a mix of 1:3 (cement:sand by volume) and as soon as this has begun to stiffen a second mortar layer is trowelled on to cover the reinforcing wires to a depth of 15mm. The surface is finished smooth with a wooden float.

After a day or so the formwork is dismantled by removing the holding bolts and by pulling out the wedges which will leave the shuttering free to be stripped away from the inside mortar wall. The sections are lifted clear of the tank to be thoroughly cleaned of any mortar or cement.

A 20cm length of 8cm diameter downpipe is built into the wall at the top of the tank to act as an overflow and the inside of the tank is plastered with mortar to fill up the corrugations. When this has hardened sufficiently a second final coat is trowelled onto the inside and finished with a wooden float.

A 5cm thick layer of mortar is next laid onto the floor of the tank and the junction with the floor and the walls built into a coving.

The floors are unreinforced and these tanks would fracture if they were moved.

Take care that the mortar does not block up the outlet pipe. Before the mortar on the floor has stiffened, form a shallow depression in the middle; this will allow the tank to be cleaned at a later date — the sediment can be brushed into the hole and cupped out (see Fig.22).

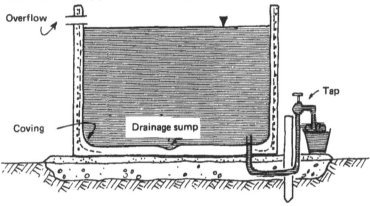

Fig.22 The completed tank

The inside of the tank is painted with a thick cement slurry to seal the tank, a small volume of water is allowed to stand in the bottom of the tank and the tank is covered and cured for seven days.

Roof

The tank is covered with sheets of 0.5mm galvanised sheeting supported on two lengths of angle iron. Alternatively, a reinforced mortar roof may be built in the ways described in Chapter 10. Building a mortar roof is not difficult but it requires extra sets of formwork.

53

Materials required for 10m³ tank with galvanised iron roof

Cement	600 kg.
Plain wire 2.5mm diameter	200m
Chicken mesh 1m wide	16m
Water pipe 20mm bore	1m
Water tap	1 No.
Overflow pipe	20cm of 8cm dia. iron or concrete pipe
Galvanised iron sheet and angle iron for roof	
Sand	1.0m³
Gravel	0.5m³

Chapter Eight

Large Open Storage Tank: 150m³ Capacity

These tanks have been built and used successfully since about 1950 and the diameters have been made as large as 25 metres to hold 450m³ of water. The walls are constructed on pre-fabricated corrugated galvanised iron formwork which is erected to form a complete circle. The cost of the formwork is a small item, coming to less than about 10% of the total cost (see Table 2) and as it will be used during the construction of many tanks its initial cost can be shared.

Design

The tank described in this chapter has a capacity of 150m³, with a height of 2m and a diameter of 10m. The walls are cast integral with a concrete foundation ring beam, and the floor of the tank is cast separate from the walls. There is no roof to the tank.

Formwork

The formwork consists of 0.6mm corrugated galvanised iron sheets rolled to a radius of 5m. The sheets can be any convenient length, but when they are bolted together they should make a circle of the correct size. If they are unstable when erected they can be propped up from the inside with lengths of wood.

The sheets are bolted into single rings with 30 x 6mm diameter veranda bolts. The bolt holes are drilled into the convex side of the corrugations (looking from the inside) at any one joint and nuts are brazed into the concave side opposite the holes. The nuts must be securely attached to the sheets to prevent turning while loosening the bolts (see Fig.23).

55

Outside face of tank

Nut brazed or welded to form

Sheets bolted together

Fig.23 Corrugated iron sheets rolled to 5 metre radius

The complete ring of shuttering can then be collapsed inwards by loosening only one joint, and by withdrawing the well greased bolts the sheets slide across one another reducing the diameter of the ring sufficiently to remove the form.

If only one or two tanks are to be built, one complete ring of corrugated iron formwork will be satisfactory; this is stripped out after it has been plastered and is lifted up to form the next ring formwork. However, if a large number of tanks of the same diameter are to be built, it is worthwhile to make up three rings for constructing the entire wall at one set up. The single rings are heavy and awkward to lift without at least 10 people to help.

Construction

The site chosen for the tank is levelled and cleared of topsoil for a radius of 6m. The inside edge of the foundation trench is marked out by attaching a 4.85m tape to a post driven into the centre of the tank site and sweeping out a circle, marking the ground every metre or so with a peg (see Fig.24).

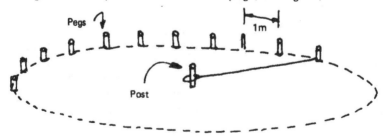

Pegs

1m

Post

Fig.24 Marking out line of foundation trench

The floor

The floor is excavated to a depth of 15cm or so inside the marked out ring and levelled with a layer of sand. The side of the excavation is formed with a packing material (fibre board, even thin bricks) and reinforcement consisting of weld mesh (20 x 20cm square, 5mm dia.) or 8mm diameter reinforcing rod, laid in place on spacing blocks. The slab may be cast in one piece without internal movement joints; a larger slab would need movement joints to accommodate shrinkage, temperature movement and settlement.

The inlet and outlet pipework should be installed before the floor is poured. Concrete made from 1:2:4 mix (by volume) is placed and compacted to make the 10cm thick floor slab. The surface of the slab is screeded level with a straight edged board and a shallow depression is scooped out in the centre; this will allow the tank to be emptied and cleaned.

The floor slab should be covered with a sheet of plastic or wet sacking for a week to allow the concrete to harden. This is especially important in hot sunny climates.

The cost of the materials to construct the floor slab is likely to make up perhaps half of the total material cost of the tank. The slab can be replaced by a layer of suitable rubber sheeting, or puddled clay, but in this case the tanks walls are constructed first.

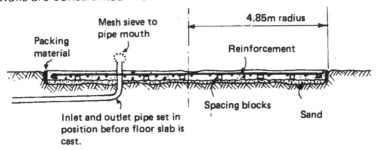

Fig.25 Casting the floor slab

The walls

A trench 30cm wide and 30cm deep is next excavated around the floor slab (see Fig.26). Take care not to undercut the floor slab.

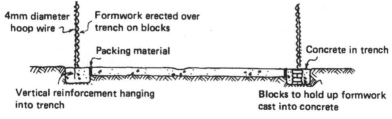

4mm diameter hoop wire

Formwork erected over trench on blocks

Packing material

Concrete in trench

Vertical reinforcement hanging into trench

Blocks to hold up formwork cast into concrete

Fig.26 Assembling the formwork

The formwork is assembled and erected on blocks or bricks that have been piled in the bottom of the ring trench. The reinforcement, pig netting or similar mesh, is drawn tightly around the formwork and overlapped about 20cm at the ends; this overlap is sewn together with soft iron tie wire.

The wire mesh is arranged to hang down into the trench where it will be concreted into place. An alternative design replaces the mesh with 6mm steel reinforcing rod at 10cm centres around the tank; this will be stronger and is likely to be cheaper to buy and transport.

The main hoop wire reinforcement of 4mm diameter is wound around the formwork on top of the mesh at the spacings given below and tightened. When a join has to be made in the hoop wire, it is preferable to fasten the ends to the pig netting or rods and give a generous overlap of 1m with the new wire. If only one ring of formwork sheeting is available the netting or steel rods are allowed to hang over into the tank.

The hoop wires are wound from the base at the following spacings: —

2 wires in each corrugation for the first twenty
1 wire in each corrugation to the top
2 wires in top corrugation

Packing material is now re-fastened around the edge of the floor slab and the ring trench is filled with 1:2:4 (cement: sand:gravel) by volume concrete. This packing material will be removed later and the slot filled with hot bitumen to provide a watertight movement joint between the floor and the wall.

The wire mesh or vertical reinforcement rods will now be

firmly concreted into the foundation beam and will help to resist cracking at the foot of the wall.

Using only a single ring of sheet formwork

With the reinforcing wire in place, a fairly dry 1:3 mortar mix (by volume) is trowelled into the corrugations covering all the wires. After 24 hours, during which time the mortar must be kept damp, two of the greased bolts making up a joint are withdrawn, the single ring is collapsed inwards and moved up into position for the next lift. If another circle of wire mesh is required it need only be overlapped 10cm horizontally and again about 10cm vertically at the ends when making fast. The hoop reinforcing is again tightly wrapped round, not forgetting the 1m of overlap, and the new lift is trowelled on as before. At the same time the corrugations on the inside of the first lift can be completely plastered up so that no reinforcement is visible from the inside.

The third lift is built in the same way, and on removal of the formwork the tank wall can be finished off by applying a 15mm mortar layer to the inside and floated to obtain a glazed watertight finish. The outside is also given a 1.5cm mortar layer. In practice the above procedure results in a wall of between 7.5-8.5cm in thickness with the reinforcing in the centre. A mortar coving is trowelled into the junction of the wall and the ring foundation beam.

When lifting the single ring to the next lift the ring is nested into the top corrugation of the previous mortar lift but in practice this does not always work due to slight irregularities in the top and bottom diametres of the formwork. The rings, however, can rest quite firmly on the top of the previous lift and require only minor adjustment in level and plumb before applying the reinforcing wire and cement mortar covering. Generally, the corrugated iron sheets come in widths of 0.66m and three separate lifts will be needed to make a 2m high wall.

Curing the tank has already been described in Chapter 6.

Having three complete rings of corrugated iron formwork will allow the tank to be built in one complete lift which will reduce the risks of cracking between improperly bonded lifts. It is also very much easier.

The packing material between the floor slab and the foundation beam is removed and hot bitumen poured in to make the movement joint (see Fig.27).

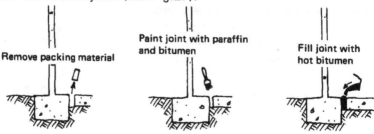

Fig.27 Making the expansion joint

Before pouring the bitumen joint seals, the joint must be completely dry and clean with no loose mortar or dust adhering to the sides or the bottom. Firstly, both faces of the joint should be painted with paraffin and then both faces should receive one coat of bitumastic paint. Immediately after painting, the joint may be filled with molten bitumen; the joint should be slightly over-filled.

Materials required for $150m^3$ tank

Walls:	Cement	2300 kg
	Sand	$5m^3$
	4mm diameter wire	1200m
	Heavy duty wire mesh, 35m rolls at 1m high	3 No. rolls
	or 6mm diameter reinforcing rods, 2.3m high	310 No. or 180 kg weight
Floor:	(Concrete slab 10cm thick)	
	Cement	2100 kg
	Sand	$5m^3$
	Gravel	$2.5m^3$
	Weld mesh	$80m^2$
	or 8mm diameter reinforcing rods (at 30cm centres)	250 kg
	Bitumastic sealer	50 litres
	Inlet, overflow and outlet pipes as needed	

These quantities may be used to estimate the on-site cost of materials needed for a tank of this size, see Table 2.

Part 3 Alternative Designs

The designs described in this part of the handbook have all been successfully built and tested. The first one describes a small rainwater jar of small capacity which is cheap because it does not require reinforcing mesh and is built onto a simple mould. The second design shows water tanks being made on a factory basis in New Zealand; the forms for these are especially fabricated from steel plate and the thickness of the walls must be carefully controlled during the trowelling. The corrugations in the self-help type formwork, in comparison, effectively ensure that the walls are of adequate thickness. Other examples show tanks made with the minimum of formwork (or no formwork at all), one of which is built onto a permanent adobe (mud and grass) wall.

These alternative designs have been included to demonstrate the flexibility of wire-reinforced cement-mortar and will perhaps encourage the reader to develop his own construction methods.

Chapter Nine

Small Jars of Unreinforced Mortar: 0.25m³, Thailand

Thailand with its wet monsoon climate has long periods of rainfall during the year. Many of the Thai people in both rural and urban areas collect the rainwater that runs off the roofs and store it in large pottery jars for domestic use. This is a traditional way of providing themselves with a domestic water supply which is both clean and convenient. During the dry periods, however, water is taken from wells or from ponds which are often contaminated causing much illness amongst the people. The water supply from the roof can be made to last much longer into the dry period simply by adding extra storage jars.

A cheap and strong water jar made from a mortar of sand and cement has been devised. These jars do not have expensive wire mesh reinforcing and can be made very large, perhaps up to 4000 litres (4m³). Jars holding 250 litres of water cost less than one tenth of the conventional clay jars to make. The users with no previous experience have been taught the skills necessary to make these jars in training periods of less than two days. These water jars enable any householder with a suitable roof to cheaply improve his water supply by his own efforts. They can also be used for grain storage.

The method illustrated in Photos 2.1 to 2.10 will make a jar of 250 litres capacity.

Photo 2 Making a small water jar: 250 litres. Thailand

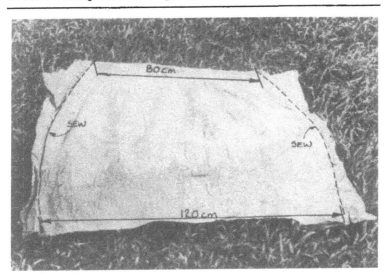

2.1 *Place two pieces of gunny cloth (hessian sacking) 125cm by 110cm together and mark out. Sew the two pieces together along the curved lines leaving the top and bottom open.*

2.2 *After sewing, turn the sack inside out.*

63

2.3 *Make a precast mortar bottom plate, 60cm in diameter and 1.5cm thick. Make the mortar from a mix of 1:2 cement:sand by volume as dry as possible consistent with easy trowelling.*

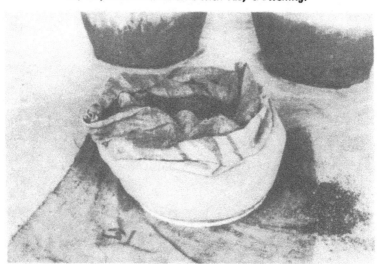

2.4 *Place the sack on the bottom plate with the smaller opening downwards and fill the space inside with paddy husk, sawdust or sand. The weight of the fill will hold the lower edge of the sack firmly on the bottom plate. Make sure that the mortar bottom plate sticks out from under the sacking.*

2.5 *When the sack is filled up, fold the top and tie it into the shape of a traditional water jar. Use a piece of wood to tap on the mould to make it round and fair.*

2.6 *Spray some water on the mould before plastering to make it damp.*

2.7 *Place a circular ring on the top of the sack to make a mould for the opening of the jar. This can be made of wood or precast mortar.*

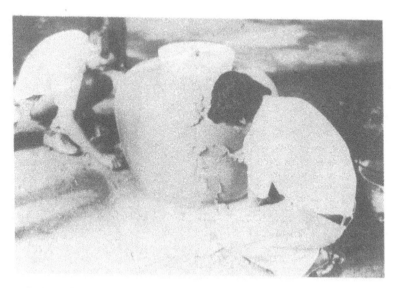

2.8 *Trowel a first layer of mortar onto the mould to a thickness of about 0.5cm.*

2.9 *Plaster the second layer of 0.5cm in the same manner as the first layer. Check the mortar layer for thickness by pushing in a nail: any weak or thin spots should be built up with an extra layer of mortar. Build up the opening.*

2.10 *Remove the contents of the gunny bag and the bag 24 hours after the jar has been made. Check the jar for any defects and correct these with mortar; the inside of the jar should also be painted with a cement slurry. Cure the jar out of sunlight and drying winds, preferably under damp sacking or plastic sheet for at least 2 weeks.*
This technique has been used with great success in Thailand and pots of up to 4000 litres (approx 1000 galls.) capacity have been made in this way.

Chapter Ten

Factory-Made Tanks
1 to 25m³ Capacity, New Zealand

Ferrocement tanks have been made commercially in New Zealand for many years and have now largely replaced the more traditional corrugated galvanised iron tanks. They are used mainly to store water for domestic and dairy purposes on the farm but they are also winning acceptance for industrial liquid storage. The cost of the smaller tanks is comparable with that of tanks made from other materials such as galvanised iron; the cost per unit volume decreases rapidly with increase in size.

Tank sizes

The tanks are constructed in various sizes, with capacities from 1m³ to 25m³, diameters from 1m to 3.6m, and heights from 1.3m to 2.9m.

Costs of constructing the various sized tanks are shown below. With specially built formwork and machine mortar mixers each tank takes from 2-5 man days to build. These costs have been obtained from an average of manufacturers' prices and reflect the relatively high wage costs in New Zealand.

Design

The water pressure in a tank full of water generates stresses in the tank that are difficult to calculate structurally. The New Zealand tanks have been designed to resist only hoop stresses and a layer of woven netting is included as nominal reinforcement; this netting in fact provides the only reinforcement at the base of the wall where it joins with the floor — the point of greatest stress. This section is thickened during construction and from information given by the manufacturers there is no evidence that cracks develop under normal loads. The only causes of failure reported have resulted from damage during delivery.

All of the tanks are built with an integral roof and a covered access hatch.

Table 4: Sizes and cost of New Zealand tanks

Capacity (m³)	Diameter (m)	Height (m)	Weight (tonne)	Price (1973) ($)NZ	Cost/m³ ($/m³)
0.9	1.20	1,30	0.25	36.0	40.0
1.8	1.55	1.30	0.30	48.0	26.5
2.7	1.85	1.30	0.45	58.0	21.5
3.6	2.00	1.45	0.80	73.0	20.5
4.5	2.00	1.95	1.25	93.0	20.5
9.0	2.90	1.95	2.10	170.0	19.0
13.5	2.90	2.60	3.0	210.0	15.5
18.0	3.65	2.60	4.00	250.0	13.8
22.5	3.65	2.90	5.00	280.0	12.5

Materials needed to make a 9m³ tank

Cement	740 kgm
Sand	1.00m³
Plain wire 4.0mm dia.	330m
Wire mesh 1m wide rolls	28m
Weld mesh for slab	7m²

These quantities are greater than those shown for a comparable tank built by self help (Chapter 7). The tanks described in this chapter have to be stronger than the self-help tanks to withstand the extra stresses produced during transportation.

Construction *(see photos 3 and 4)*

The tanks are constructed on special fabricated steel formwork which is quickly erected (see Fig.28) or on to temporary timber formwork. Usually the floor of the tank is cast first; this is reinforced with welded steel mesh made from 8mm diameter rods at 20cm centres and given a thickness varying from 6cm for the small tanks to 10cm for the larger tanks. Loops of 8mm steel are allowed to project from the sides of the base to allow for easy handling; this also reduces the stresses that will be set up in the tanks as they are being lifted

Photo 3 Construction details of commercial tanks (New Zealand).

3.1 *The inside former, which provides the tank shape, is usually a series of vertical planks supported by a steel frame. These will be sheathed in sheet steel before applying the plaster.*

3.2 *After applying the initial plaster the reinforcing is placed. This consists of a layer of netting and a continuous spiral of wire.*

Photo 4 Construction details of commercial tanks (New Zealand) cont.

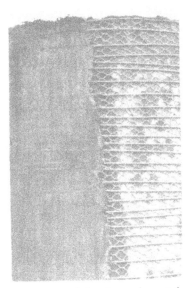

4.1 *The reinforcing wire and mesh become embedded in the plaster build-up.*

4.2 *A corner of a ferrocement manufacturer's yard. The white-painted tanks towards the left are complete and ready for delivery. The tanks on the right are at various stages of roof construction. The elliptical septic tanks in the foreground are also manufactured by the ferrocement technique.*

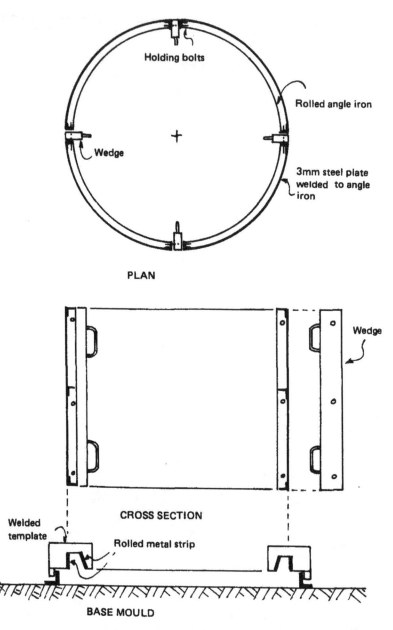

Holding bolts

Rolled angle iron

Wedge

3mm steel plate
welded to angle
iron

PLAN

CROSS SECTION

Wedge

Welded
template

Rolled metal strip

BASE MOULD

Fig 28 Commercial tank formwork

or winched. Experience over many years with these tanks indicates that the only failures that have occurred have been due to rough handling during delivery. A strip of chicken wire is also cast into the sides of the floor and is bent up into the walls (see Fig.29).

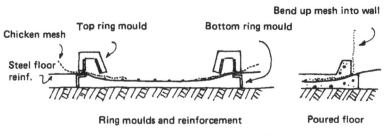

Fig.29 Casting the base slab

When the floor slab has been cast the formwork is erected and the chicken wire folded up against the shuttering. A layer of chicken wire or weld mesh made from 2mm wire at 5cm centres is wrapped around the tank to cover the shuttering from top to bottom (see Fig.30).

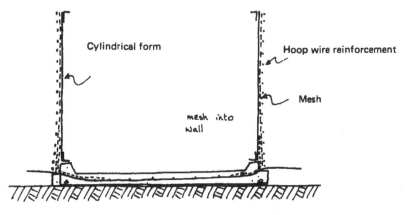

Fig.30 Assembling the formwork and reinforcement

The main reinforcement, 4mm diameter straight wire, is wrapped tightly around the tank in a spiral with a 5cm gap between the wires. Theoretically this gap should be much smaller at the bottom of the tank than at the top to take the greater stresses, but in practice the spacing is left constant.

73

This prevents mistakes during construction and does not add appreciably to the total costs. The same spacing is often used on all of the tanks, both small and large.

The first mortar layer (1:3 cement:sand by volume) is trowelled onto the tank 1cm thick and given 24 hours to harden. A second layer of mortar is then trowelled on and finished smooth with a plasterer's float; this is also given 24 hours to harden.

The formwork is now carefully stripped and removed from inside the tank and a third layer of mortar is trowelled on the inside to completely cover up the reinforcement. A thick unreinforced coving is added to strengthen the joint between the walls and the floor of the tank.

Finally, the roof is built onto the tank by laying mortar onto shaped formwork which is propped from underneath. The roof is reinforced with two layers of wire mesh which is tied onto the mesh protruding up from the walls (see Fig.31).

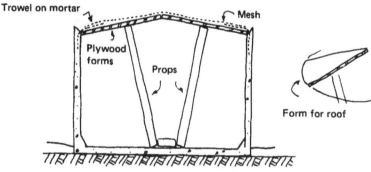

Fig.31 Constructing the roof

A prefabricated angle-iron frame is set into the wire mesh to provide formwork for an access hatch into the finished tank. This is removed after the mortar has set (see Fig.32).

Mortar is trowelled on in a 3cm layer and allowed to cure for three days. When it is strong enough the roof and access hatch formwork is stripped and a layer of mortar trowelled onto the underside of the tank roof.

The tank is finally painted inside with a coat of cement and water slurry, a small volume of water is allowed to stand in the bottom of the tank and the tank is covered and cured for at least seven days for the mortar to harden properly.

74

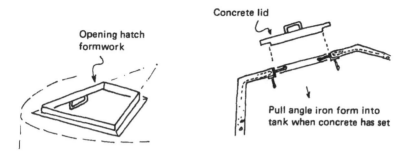

Fig.32 Making the access hatch

Transporting the tanks

The factory made tanks of less than 25m³ capacity are light enough to be carried by lorry. They are taken to the prepared site and joined directly to the necessary pipe connections; tanks of larger capacity (greater than 25m³) are usually built on the site. The smaller, lighter tanks are lifted onto and off-loaded from a truck with a truck mounted hoist.

The larger tanks are winched onto the truck with a sling. The first step is to jack one edge of the tank clear of the ground. The truck is then so positioned that a pair of steel runners resting on its carrying platform can be placed under the tank to form a ramp. A wire rope sling is fitted round the tank which is then drawn up the ramp by a winch mounted on the truck. Steel pipes are used as rollers when moving the tank.

For unloading, the platform of the truck is raised slightly and the tank slides down the ramp. The steel pipes are again used as rollers and the downward movement is controlled by the winch.

Chapter Eleven

Demonstration Tank Built without Formwork: 6m³ Capacity, U.K.

This tank was constructed for demonstration purposes in order to investigate the feasibility of constructing a tank without any shuttering. The reinforcement form was built from welded mesh and covered with two layers of chicken wire. Mortar was applied from the outside and forced through the netting to the inside: it would have been better to apply the mortar from the inside, but this would have made the construction work more difficult as all of the mortar would have had to be lifted in through the roof.

From the experience gained in this project we do not recommend this method of construction, although it is similar to the more conventional techniques of ferrocement. The extra cost of the welded and woven mesh would have paid for a cheap shutter and, because of the thin gauge of weld mesh used, the mortar core was flexible and moved during plastering causing the mortar core to slump. The design has been included because some of the features of construction are of similar design to others.

Tank size

The tank was constructed with a diameter of 2m, a height of 2m and a wall thickness of 3cm. The capacity of the tank was 6.3m³

Design

The tank was designed to be constructed without any formwork; the preliminary calculations indicated that the maximum loads that the tank would have to take would not overcome the strength of even unreinforced mortar. The reinforcement was therefore included for ease of construction and to prevent shrinkage during curing, rather than for strength.

Construction *(see photos 5.1 to 5.11)*

A circle of 3m diameter was marked on the ground and cleared of topsoil; the excavation was backfilled with a layer of sand and gravel. A peg was driven into the gravel through the centre of the foundation and a circle of 1m radius traced out around the peg. This marked the outside face of the tank and a row of disused bricks was placed to act as temporary formwork to the concrete floor. A layer of weld mesh was laid across the circle and strips of chicken wire tied around the edges. The weld mesh and chicken mesh were bent up in a circle over the inside edge of the bricks, and a layer of cardboard fixed to the inside, allowing the coving on the inside of the tank to be built up. Alternatively, the reinforcement may be topped by a second layer of bricks and bent up at a later date when the floor slab has hardened (see Fig.33).

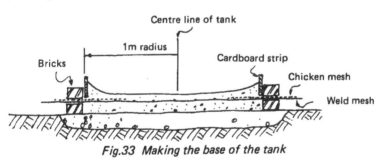

Fig.33 Making the base of the tank

When the floor slab had hardened, a 3m high section of weld mesh was erected in the shape of a cylinder on the floor slab, the floor mesh and weld mesh was bent up and tied to the cylinder with soft iron tie wire. Two layers of chicken wire were then wrapped around the weld mesh formwork and tied securely into place (see Fig.34).

A mortar core wall of 1:3 (cement:sand by volume) was then trowelled onto the reinforcement. One man stood outside the tank and pushed a trowel full of mortar through the reinforcement to a man standing on the inside of the tank

Photo 5 Constructing experimental tank on weld mesh frame (UK).

5.1 *Note bricks acting as temporary formwork*

5.2 *Tread mortar into corner.*

5.3 *Finished floor*

5.4 *Bend up reinforcement*

5.5 *Build up weld mesh for walls*

5.6 *Tie on chicken wire*

5.7 *Plaster on core wall*

5.8 *Plaster outer layer*

5.9 *Formwork for roof.*

5.10 *Building up the roof. Note the roof shutters.*

5.11 *Finished tank painted white.*

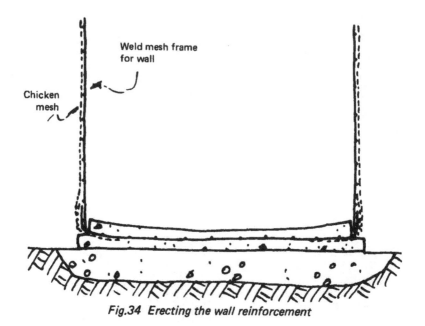

Fig.34 Erecting the wall reinforcement

also holding a trowel. This process proved time consuming and difficult, as during the trowelling the reinforcement tended to shake and move causing the mortar to slump down. This could perhaps have been avoided by using one or more extra layers of chicken mesh but at extra cost. Great care was taken to make sure that the mortar penetrated into the reinforcement at the junction with the floor slab.

The core wall was given a day to harden and 1cm thick layers of mortar were trowelled onto both the inside and outside faces of the tank. The tank walls were given a week to cure under a cover of black plastic sheeting.

The roof was next constructed by bending down the weld mesh and fastening on top of this a series of sections to form a core with a 30cm diameter hole in the centre. Chicken wire was attached, shuttering was propped up from beneath and the roof was cast in mortar (see Fig.35).

After several days, the shuttering was stripped out and the inside of the tank given a wash of thick cement slurry. The tank was again allowed to cure for a week under cover and the outside of the tank painted white.

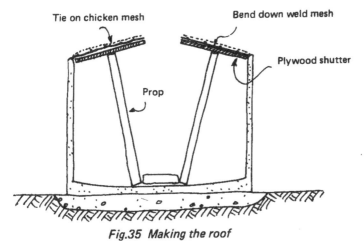

Tie on chicken mesh

Bend down weld mesh

Plywood shutter

Prop

Fig.35 Making the roof

Materials required for construction

Cement	400 kg
Weld mesh, 15cm mesh, 5mm diameter, 4m x 2m sheets	2 No.
Woven mesh, 2cm openings, 1m high rolls 30m long	1 No.
Sand	$1.0m^3$
Timber and plywood for roof shutters	

Chapter Twelve

Traditional Adobe Granary Bins, Ferrocement Lined, Converted to Water Tanks: 10m³ Capacity, Mali

The Dogon people of Mali, living in the Sahelian drought zone with a rainfall of only about 40cm per year, suffer greatly from water shortages during the dry season. The method of storing water described in this chapter was devised to provide a cheap tank for water collected from the flat roofs of the houses. Rain is of such importance to the Dogon people that they have built many myths and customs around the coming of the rainy season. The water tanks, which consist of traditional grain bins, lined with a thin layer of reinforced mortar, are readily acceptable to the users and fit well into their social and cultural visions of life. The bins already exist outside each group of houses and can be adapted with the minimum of work and skills on the part of the users (see Fig.36).

TOTAL ROOF AREA: 80 SQUARE METRES

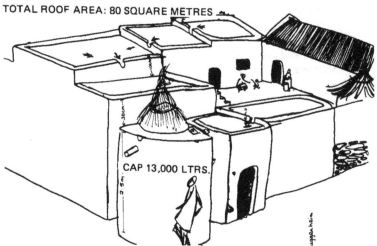

CAP 13,000 LTRS.

Fig.36 Rainwater tank outside houses, Mali

Design

The tanks are built in a way that could not be simpler, consisting of traditional grain bins lined with a layer of chicken wire and mortar. The cylindrical adobe walls provide the formwork which supports the mortar, and if covered at the top with plastic sheeting allow the mortar to cure safely. The bottom of the tank is built up into a basin shape using a soil-cement mix; this avoids corners in the tank and reduces the risks of cracking under load. The single 1cm thick layer of mortar will not support the weight of water in a full tank on its own but most of the load will be taken by the thick adobe walling which is reinforced with grass fibres. The tanks are about 2.4 metres high and 2.6 metres in diameter. The roof is flat with a removable dome for access; this prevents rubbish from falling into the tank and cuts down evaporation losses.

There will be difficulties when the adobe walling which provides a permanent support begins to erode in the rain. If the users do not keep this in good repair the tank is likely to crack and leak.

Construction *(see photos 6 and 7)*

If a new storage tank is to be built the site must be cleared and an adobe bin built up in the traditional way. The bottom of the bin is rounded into a basin shape using a soil-cement mixture (about 1:10 cement:soil by volume) and the chicken mesh stapled around the inside of the bin. Cement mortar of 1:3 mix (cement:sand by volume) is trowelled on by hand in a single layer 1cm or so thick over the walls and floor. A small amount of water is stood in a bucket at the bottom of the tank to keep the air moist and the top of the tank covered with a layer of plastic sheeting. This effectively seals the tank and allows the mortar to cure.

Converting an existing granary into a water tank follows exactly the same procedure.

The roof of the tank is built in the traditional way by constructing a timber frame laid flat over the top of the tank, covering this with branches, then building up a layer of adobe. The traditional access through a removable thatch lid is replaced by a permanent adobe cover.

*Photo 6 Ferrocement-lined adobe granary bin under construction.
Mali*

Photo 7 Building adobe tank, lined with ferrocement. Mali

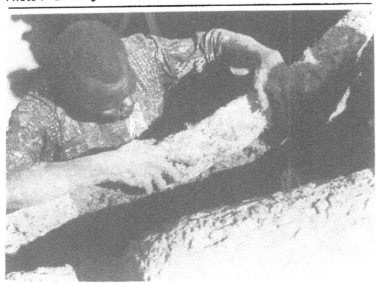

7.1 *Building the adobe walling*

7.2 *Outside view*

7.3 *Lining tank with chicken mesh and mortar*

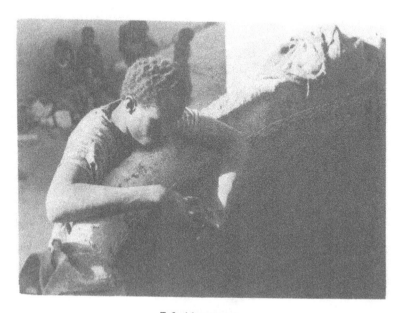

7.4 *Lined tank*

7.5 *Building the roof*

7.6 *Building the roof, bottom right*

The roof to the tank can be considerably improved by replacing this type of structure by a slab of reinforced concrete with a lid for access (see Fig.37).

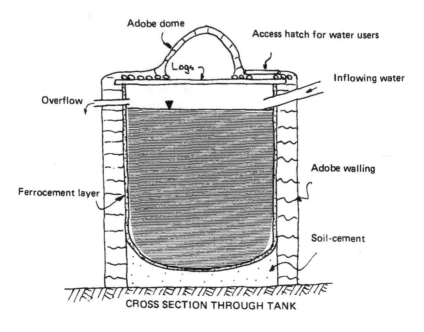

Adobe dome

Access hatch for water users

Logs

Inflowing water

Overflow

Adobe walling

Ferrocement layer

Soil-cement

CROSS SECTION THROUGH TANK

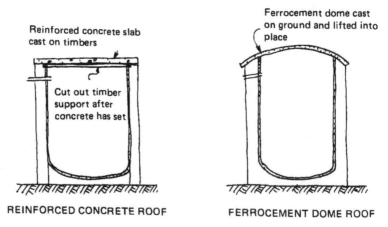

Reinforced concrete slab cast on timbers

Cut out timber support after concrete has set

Ferrocement dome cast on ground and lifted into place

REINFORCED CONCRETE ROOF

FERROCEMENT DOME ROOF

Fig.37 Ferrocement-lined adobe tanks

Alternatively, the roof may be built from a ferrocement dome reinforced with straight wire. This may be built on the ground on a simple mould built up from earth and covered with a plastic sheet, or directly on to the tank over temporary formwork built from timber and sticks.

The domes are a naturally strong shape ideally suited for ferrocement.

Water is taken from these tanks from a hole in the roof (see Photo 8). This obliges the users to climb up for water which will encourage rationing but is likely to increase the risk of accidental contamination. For a newly constructed tank the more usual pipe may be built into the floor with a tap on the outside; unless the users are careful, however the tap may be left running and the water in the tank will then be wasted.

Materials required (for a 13m³ capacity tank)

Cement	450 kg
Sand	0.7m³
Chicken mesh 1m wide	15m
Polythene sheeting	5m²

Photo 8 Using the tanks

8.1 *Lifting cover off tank*   **8.2** *Lifting water out of tank*

91

Chapter Thirteen

Roofed Tank with makeshift Formwork: 40m³ Capacity, Rhodesia

These tanks have been developed for construction in rural areas and their main feature is the use of cheap plywood, sheet iron, or even woven stick formwork. The sheets are held in place by vertical angle iron sections set into the foundations to make a circle. The reinforcing wires are tied to the outside of the angle-iron sections which are embedded into the wall structure during plastering (see Photo 9).

The tanks are built with a diameter of 5m and a depth of 2m. They hold 40m³ of water. The final wall thickness is approximately 5cm built up from a 1:3 (cement:sand by volume) mortar.

Design

The tanks have been designed to take hoop stresses only and no vertical reinforcement is included. The vertical angle iron will resist most of the force of the water tending to push the base of the wall outwards. It would be expected that loading would concentrate at the base of the angle irons to cause cracking, but long field experience has demonstrated that this does not occur. The tank walls are carried on a foundation and are not continuous with the floor of the tank; the floor can therefore be made from puddled clay.

Construction

A suitable site on firm ground is chosen and the topsoil cleared to form a circle of 6m diameter; the centre of the circle is marked with a peg.

The wall foundation trench is marked on the ground by attaching a tape measure to the centre peg, tracing out a

Photo 9 Roofed tank, Rhodesia

9.1 *Marking out and excavating foundations*

9.3 *Tying on reinforcing wire*

9.2 *Backfilling foundations with concrete and placing angle iron*

9.4 *Mixing the mortar.*

9.5 *Laying on the mortar.*

9.6 *Making the roof.*

9.7 *Curing walls*

9.8 *Finished tank with inlet.*

9.9 *Catchment area for tank with wire fence.*

radius of 2.5m on the ground. A trench 30 x 30cm is dug. A similar trench for any pipework is also excavated and the pipework installed.

A concrete mix of 1:2½:5 (by volume) is then placed in all foundations; it should be well compacted for maximum density and finished off with a level surface.

Twenty four lengths of 2.3m long angle iron, 30 x 30 x 4mm sections are cast into the wet foundation concrete at 65cm centres to a depth of 30cm. These must be placed truly vertical and supported whilst the concrete sets as they will determine the shape of the tank; the concrete should be given several days to harden before the walls are constructed.

'Straight' galvanised steel wire, 4mm diameter, is now wrapped around the ring of vertical angle iron supports, starting from the bottom at 5cm spacings. At each angle iron each strand of wire is tied with soft iron wire. Three extra strands are wrapped around the top of the tank.

The formwork to make the tanks consists of sheets of thin plywood or flat galvanised iron sheeting. When the reinforcing wire has been attached up the full height of the tank the formwork sheeting is fastened inside the ring of angle iron supports, and tied to them with soft iron wire. The walls are now ready to be plastered.

A mortar layer is trowelled 1cm thick onto the outside of the tank, given 24 hours to harden and a second coat applied and finished with a plasterers' steel float.

Some tank builders, who have limited quantities of formwork sheeting, use one or two sheets only and plaster the tank walls in sections with complete success. In this case the sheeting is left in place until the mortar has hardened and is strong enough to stand up on its own. The formwork is then stripped from the tank and moved to an adjacent panel. We do not recommend this economy, however, as complete sets of self supporting shuttering make construction very much simpler.

The formwork is finally stripped from the inside of the tank and a third mortar layer 1cm thick is applied to the inside of the tank. Finally, a fourth mortar layer is trowelled onto the inside of the tank and the tank is covered and allowed to cure.

The roof

If a roof is to be built over the tank, it is given a 50cm diameter access hatch at the centre to allow access for cleaning. The roof reinforcement consists of steel rods, 8mm in diameter which are placed at 32.5cm centres around the top of the tank, giving alternate rods which are either 3m long and fastened to the angle iron in the walls, or 2m long and fastened to the hoop wire, before the walls are plastered. The rods on the angle iron supports are bent down to form a canopy then tied temporarily to a prop erected in the centre of the tank; this will leave an opening for the access hatch. The shorter rods attached to the hoop wires stop short of the centre support. 4mm diameter wire is wound in a spiral over the top of the tank at 10cm spacings and tied to each radial reinforcement rod with soft iron tie wire.

One section of the reinforcement is often left loose to allow the sheet formwork of the walls to be lifted out. At this stage, the walls are ready to be plastered (see Fig.38).

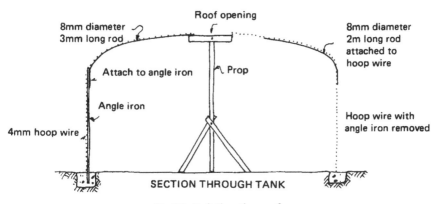

Fig.38 Building the roof

Plastering the roof is carried out after the walls have been plastered in stages starting at the sides of the roof (if there is insufficient sheet formwork). The sheets are propped up from below hard against the reinforcement and a 3cm layer of mortar trowelled onto the top. When this has hardened after 2 or 3 days, the sheet is stripped and moved to an adjacent, unplastered part of the roof. This is plastered in

After the roof has been completed the floor of the tank is laid with a 7cm thick reinforced concrete layer. This part of the work can be carried out before the roof and walls are constructed in the way described in Chapter 8; alternatively, an access hatch at the edge of the roof may be unplastered through which the concrete can be lowered.

Materials needed for construction

Cement	2010 kg
Angle iron 30 x 30 x 4mm,	
24 No. @ 2.3m long	55m
Straight wire 4mm diameter	900m
8mm diameter reinforcing rod,	
24 at 3m long, 24 at 2m long	120m or 65 kg

Chapter Fourteen

Open Tank with Makeshift Formwork: 150m³ Capacity, U.S.A.

These tanks have been built for over 20 years by farmers in Arizona, U.S.A., to provide cheap and durable water storage for their cattle. They are built by one man in 90-100 hours of work, using powered mixing equipment and a pick-up truck to collect the aggregate. The tank is reinforced by two layers of weld mesh covered inside and outside with woven chicken wire. Metal sheets are wired to the outside of the stiff wall of reinforcement to provide a temporary back-up form as the mortar is plastered on from the inside. This method of tank building is relatively expensive due to the cost of the mesh reinforcement. The tanks are used mainly for water storage for cattle and are usually filled from wind powered pumps (see Photo 10).

The tanks are built with various capacities, from 20-150m³ with diameters from 2.5 to 10 metres and a depth of 2m. The one described in this section has a capcity of 150m³

Design
The tanks have been designed as conventional reinforced con- crete and the chicken wire is assumed to contribute little towards the strength of the tank. The maximum size of tank that will stand up without cracking has been calculated to be 10m diameter and several tanks of this size have been used in the arid Arizona climate without fracturing. The tanks are constructed with the floor and the walls joined joined to- gether without reinforcement at the joint; experience has shown that the joint does not crack and allow leakage. The floor is unreinforced. The tanks are located so that inlet to the tank fills the tank by gravity; when the tank is full water overflows back into the well beneath the windmill.

Photo 10 Building the tanks, USA

10.1 *Cheap shuttering from flattened oildrums*

10.2 *Applying mortar from inside the tank*

10.3 *Removing shuttering*

10.4 *Tank with cattle watering trough*

101

10.5 *Water pumping into tank*

Construction

The site area is cleared of surface vegetation and the line of the tank walls is marked out on the ground with a 5m radius from the centre peg.

The rolls of weld mesh for the 2m high inside and outside layers from 5mm steel wire, are unrolled on the ground and joined to give the desired length of 31.5m. The outside layer is folded down 30cm at the top and the inside hoop is folded up 40cm.

The mesh layers are re-rolled, carried to the tank site, opened up and tied together to form a wall of double layered mesh on the marked out circle (see Fig.39). The mesh is held

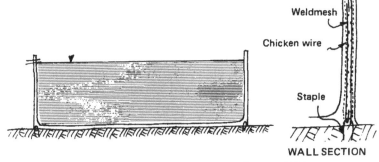

Weldmesh

Chicken wire

Staple

WALL SECTION

Fig.39 Construction of 150m³ tank, USA

to the ground by 30cm long 8mm steel rod staples bent through the mesh and hammered into the ground. The chicken mesh is added next with a layer on the inside and outside of the tank; this is fastened securely in place with soft iron tie wire. Care is taken that the overlaps to the weld mesh are staggered as the extra thickness at the overlap makes construction very difficult. The chicken wire and weld mesh are tied together beginning at one end of the mesh and continuing to the other end; this prevents the wire mesh from creeping during tying and causing bulges.

When these tanks are used primarily for stock watering the outlet pipe is pivoted to allow the pipe to be adjusted to the required depth below water level. If the water control float on the cattle trough fails then only a small fraction of the water in the tanks is lost. With the pipework fixed in position the concreting can begin.

103

Mortar made from 0.2:1:3 (lime:cement:sand by volume), is built up around the base of the wall reinforcement and allowed to harden. The hydrated lime is added to the cement to improve the workability of the mortar. Metal sheets 1m wide are fastened around the outside of the reinforcing cage to provide a back-up when the mortar is plastered on from the inside. A hole is made through one of the sheets and the chicken wire cut away to allow mortar to be poured through to the inside of the tank. This panel is the last to be completed. A mortar layer 2cm thick is then trowelled on and pushed through the reinforcement to the back-up steel sheets; the chicken mesh holds this thick layer onto the shutter without slumping.

Immediately after the core wall has stiffened (about 6-12 hours, depending on the climate), the sheets are stripped from the base of the tank and fastened to the top section which is again plastered from the inside.

The mortar 'core' wall is covered with black plastic or wet sacking as soon as it has been plastered to reduce water evaporation during curing. After initial hardening the metal sheeting is stripped, and 1cm layers of mortar are applied to the inside and the outside of the tank to give a final wall thickness of 6.5cm. If the core wall has dried out it is splashed with water before plastering.

The tank floor is covered with a 1:2:4 (cement:sand:gravel by volume) concrete to a depth of 10cm. The base of the wall on the inside and the outside is finally built up to a thick unreinforced coving, the mixer shute hole is patched up, and the tank is given 7 days to cure under wet sacking or black plastic.

Materials needed for 150m³ tank

Walls:	Cement	2300 kg
	Lime	450 kg
	Sand	5m³
	Weld mesh, 2m high	62m
	Woven mesh, 2m high	62m
Floor:	Cement	1850kg
	Sand	2.6m³
	Gravel	5.1m³
	Asphalt paint	45 litres

In this design, the inside of the tank is painted with a layer of black emulsified asphalt paint to reduce the risks of leakage through fine cracks.

Chapter Fifteen
Calculated Stresses in Thin-Walled Cylindrical Water Tanks

Summary
The design calculations presented in this chapter have been prepared from an idealised conception of a thin walled tank which does not occur in practice. However, they have been included to allow a tentative assessment to be given of the strength of the tanks and their safety margin against failure.

The calculations show that the greatest area of stress for all of the tanks occurs across a horizontal plane on the inside face of the tank at the junction of the wall and the floor, which in the larger tanks exceeds the maximum permissible tensile stress for cement mortar. This stress is greater than the stress that has to be resisted by the hoop reinforcing wire. Most of the construction designs described in Part 3 have only minimal reinforcement in a vertical direction and this is a major design weakness.

Design assumptions for calculations
  i. The tanks are assumed to be made of a uniform, homo-genous, elastic material, with an elastic modulus the same in both horizontal and vertical directions. Poissons ratio for reinforced mortar is taken to be zero.
  ii. The tanks have a cylindrical wall with a flat floor; the walls and the floor are composite, and the design stresses are assumed to be low enough to prevent a plastic hinge developing at the inside corner of the wall; no account has been taken of creep which would relieve some stress in the mortar.
  iii. The liquid in the tank is water and fills the tanks to the top of the walls.
  iv. All calculations have been taken from the publication,

Reinforced Concrete Reservoirs and Tanks, G.P. Manning, Cement and Concrete Association, 1972. The values of the dimensionless term $(H \times RT^{-\frac{1}{4}})$ are beyond the range shown by Manning:—
where: H = tank height
R = tank radius
T = wall thickness
v. The dimensions and wall thicknesses of the tanks have been chosen to be representative of the tanks described in Part 2.

Table 5 Calculated stresses in thin walled, cylindrical tanks

Tank dia.	Wall thickness	Capacity	Maximum hoop stress	Position of max. hoop stress	Max. bending stress on inside face	Shear stress at base
m	cm	m^3	N/mm^2	$\left(\dfrac{h}{H}\right)$	N/mm^2	N/mm^2
2.50	3.0	9.0	0.73	0.20	1.32	0.09
5.0	3.0	40.0	1.26	0.24	2.34	0.16
5.0	5.0	40.0	0.65	0.29	1.42	0.11
10.0	6.5	150.0	0.88	0.42	2.06	0.13
20.0	8.0	630.0	0.86	0.55	2.26	0.13

Symbols: H = tank height (= 2.0m)
R = tank radius m
T = wall thickness cm.
h = distance from base

Comments: In all cases, the maximum tensile stress occurs on the inside face at the junction of the wall and the base and is over double the maximum hoop stress.
The calculated stresses in both horizontal and vertical directions are low for the tanks less than 10m diameter, and the mortar may be expected to withstand this load without cracking. The 10m and 20m diameter tanks develop tensile stresses at the base greater than that normally allowed for concrete and the addition of vertical wire reinforcement here will probably make a major improvement to the design.
The shear stresses at the base of the walls are low and present no difficulties in design.

Most of the tanks described in Part 2 have hoop wire reinforcement with a layer of wire mesh on the inside face of the tanks to aid construction and prevent shrinkage cracking. The condition of equal elastic modulus in a horizontal and vertical direction will therefore not occur in these cases, although the wire mesh will contribute substantially to withstanding the load stresses on the inside face of the tank. What is likely to happen here is the development of a plastic hinge on the inside face of the tank at the position of maximum stress, i.e. at the base; the wall will then deflect outward and throw a greater load onto the hoop steel. The cracks developed at the plastic hinge, if they open wide enough, could allow water to reach the reinforcing wire and cause corrosion. This would lead to the eventual failure of the tanks although the cracks can be repaired simply by painting the joint with bitumen. A large number of these tanks have already been built and used successfully for over 25 years and this potential cracking does not seem to be critical in practice.

Chapter Sixteen

Catching and Using Rainwater from the Roof

The amount of rain that falls and may be caught on a suitable roof depends on the climate of the area. Rainfall amounts and hence catchment water, will vary between seasons and years, and rainfall records will allow only a very approximate estimate to be made of the possible quantities of water that will be available from the roof. Deciding whether or not a catchment tank is worthwhile, and if so, its optimum size is therefore largely a matter of local judgement and experience, especially where there are other water sources that could be developed. A trial tank built for a typical house in an area will give a useful indication of the practicality of a more extensive tank construction programme and will also encourage the local people's interest. The following steps have been included in this chapter to indicate and illustrate the various factors that affect tank size and use; they will give only a rough guide to the size of tank needed.

a. Rainfall, potential run-off from the roof

The rain that falls on the roof runs off into a gutter and is led into a storage tank (see Fig.40). As 1mm rainfall gives $1 l/m^2$ of horizontal surface, therefore total run off = R x A litres, where A = Plan area of roof (m^2) and R = Rainfall (mm).

Example

Rainfall	= 50mm
Run-off/m^2 roof	= $50 l/m^2$
Roof plan area	= $30 m^2$
Total run-off from roof	= 30 x 50 = 1500l

b. Potential supply month by month

The rainfall is caught on the roof; it drains off through the guttering and is stored in the tank. As the run-off is filling the

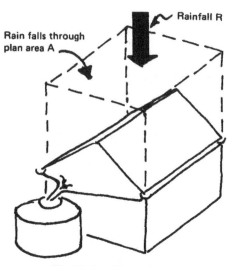

Rain falls through plan area A

Rainfall R

Fig.40 Catching rainwater

tank water is also being taken out for use. The size of the tank must be made large enough to provide enough storage, if possible, to last through any dry periods. In order to find the cumulative run-off from the roof month by month over the year, we assume at first that the water in the tank is not being used (see Fig.41).

c. Rate at which water can be used from the tank

The amount of water used each day will vary considerably depending whether the users waste or ration the water, but a value of about 15-20 litres per person per day is often quoted. In practice, most householders will use the water copiously during the rainy periods, especially if the tank is overflowing, and will ration it or find another source during dry periods.

Water is being taken from the tank for use at the same time as it is being collected. A roof of given size can collect only a fixed amount of water and hence yield only a fixed supply. The storage volume needed to provide the maximum steady rate of water usage can be determined by plotting the cumulative run-off from Fig.41, by drawing in a line representing cumulative water use touching the first curve at only one point, allowing the storage volume and the potential daily yield to be measured. In Fig.42, it has been assumed

110

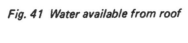

Fig. 41 Water available from roof

Rainfall/month (mm)

Dry Season

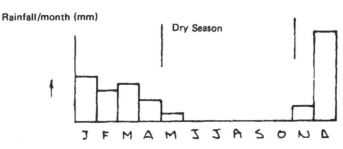

J F M A M J J A S O N D

i. Plot average monthly rainfall figures (Rmm) from all available records.

Total runoff/month (l)

ii. Calculate run-off for each month (= R x A) litres.

Total runoff added cumulatively (l)

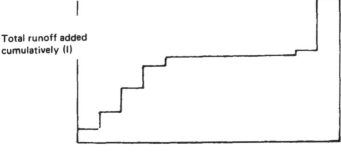

iii. Add monthly figures cumulatively (litres)

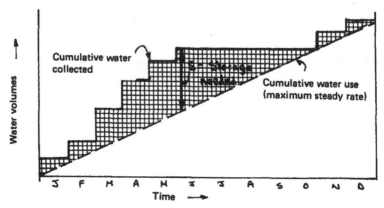

Cumulative water collected

Cumulative water use (maximum steady rate)

Water volumes

J F M A M J J A S O N D

Time

Fig. 42 Estimating tank size

that the tank is filled up to capacity only once during the year.

In the real world the climate is unlikely to be uniform enough to allow these steps to give more than a rough indication of tank size. In very dry areas, the roof will usually be too small to catch the volume of water needed for the whole year, but even in this case the stored water may still provide a useful, uncontaminated and controllable water source for at least part of the year.

Chapter Seventeen
Sources of Further Information

Contacts

Intermediate Technology Development Group,
9 King Street,
Covent Garden,
London WC2,
United Kingdom

ITDG's Ferrocement Panel members are able to answer queries about the use of ferrocement in water storage applications. More detailed enquiries should be directed to the relevant sources quoted below.

Friends Rural Training Centre,
Hlekweni,
P.O. Box 708,
Bulawayo,
Rhodesia

The founder of this centre for rural artisans, Ray Henson, has been making the self-help tanks described in Chapter 7 for many years.

Co-operative Extension Services & Agricultural
    Experimental Station,
College of Agriculture,
The University of Arizona,
Tucson,
Arizona,
U.S.A.

The tank described in Chapter 14 was devised by this Extension Service for stock farmers. They provide helpful advice.

Ian Crayford,
Honiara Marina and Shipyard Company,
P.O. Box 144,
Honiara,
Solomon Islands

This company has had wide experience in ferrocement boat building and is turning its attention to the use of ferrocement in waterpipe and water tank construction.

Dr G.L. Bowen,
Senior Lecturer in Civil Engineering,
University of Aukland,
School of Engineering,
Private Bay,
Aukland,
New Zealand

Dr Bowen is able to answer enquiries on behalf of the New Zealand Ferrocement Marine Association. He will be able to direct any enquiries on the tanks described in Chapter 10 to the relevant source.

Cement and Concrete Association,
Wexham Springs,
Slough,
Bucks

The C & CA published many 'man on the job' leaflets on concrete and concrete use, much of which is relevant to ferrocement applications. These basic instruction leaflets are of great use in training programmes.

Howard Liddel,
School of Architecture,
Hull Regional College of Art,
Brunswick Avenue,
Kingston-upon-Hull,
HU2 9BT
United Kingdom

Howard Liddel has been directing a research project with his students on the possibilities of providing water supplies from the house roof. His department is carrying out field studies on roof catchment water supplies.

E.A.J. Canessa,
c/o Public Works Department,
23 John Mackintosh Square,
Gibraltar
The authorities in Gibraltar insist that each house uses its roof to collect rainwater for domestic use. Further information on this may be requested from Mr Canessa.

Department of Civil Engineering,
Asian Institute of Technology,
Bangkok,
Thailand
AIT organised a workshop on ferrocement in December, 1974. They are investigating the potential of ferrocement applications in developing countries for a broad variety of uses. The 'International Ferrocement Information Centre' has just been established at the AIT to act as a clearing house for information on ferrocement and related materials. They will send further information on request.

References

Tanks for water from roofs in Botswana and S.W. Rhodesia,
R. Henson, 1972, Friends Rural Training Centre, Hlekweni,
P.O. Box 708, Bulawayo, Rhodesia. 2pp.
   A basic description of the tank in Chapter 7. The tanks
have been developed and designed to allow their safe construction by self help. The tanks have been built for a number
of years with complete success.

'Reinforced plaster reservoirs', Bulletin 2547, *Rhodesia
Agricultural Journal,* Volume 67, No.3, May/June, 1970,
Ministry of Agriculture, Agricultural Research Centre, 5th
Street, P.O. Box 8108, Salisbury, Rhodesia. 6pp.
   The tank described in Chapter 8 was described in this
article. A design method is presented for different sized
tanks to determine the amount of reinforcement required
for tanks of different depths; the method does not consider
the strength contribution of the wall and floor joint.

'Water jars from cement mortar', *Appropriate Technology,*
Volume 2, No.2, August, 1975 from Intermediate Technology
Publications Ltd., 9 King Street, London WC2E 8HJ.
   A description of the water jar design described in Chapter
9. Further information may be obtained by writing to:—
Mr Opas Phromratanapongse,
The Siam Cement Co Ltd.,
P.O. Box 572,
Bangkok,
Thailand

Ferro cement tanks and utility buildings. An undated report prepared by the New Zealand Portland Cement Association, P.O. Box 2792, Wellington, New Zealand. 8pp.

This association is also preparing a Code of Practice on concrete tank construction. The commercial tanks are described in Chapter 10.

Fighting the Drought, Hans Guggenheim, The Wunderman Foundation, 575 Madison Avenue, New York, NY 10022, U.S.A., 1974.

The ferrocement lined adobe tanks described in Chapter 12 were devised for use in the drought zone of Mali to be acceptable to the users by fitting in with local beliefs and customs. The tanks are an adaption of traditional granaries which are built alongside the houses.

Circular Reinforced Plaster Reservoirs, Rhodesia, Council of Social Services, Ministry of Labour & Social Welfare, Private Bag 7707, Causeway, Salisbury, Rhodesia. Undated report. 5pp.

A description of the tank from Chapter 13.

How to make a plastered concrete water storage tank, Bulletin A-41, Co-operative Extension Service and Agricultural Experimental Station, University of Arizona College of Agriculture, Tuscon, Arizona, U.S.A., 1965. 12pp.

A description of the tank from Chapter 14.

Ferrocement: applications in developing countries, National Academy of Sciences, 2101 Constitution Avenue, N.W., Washington D.C. 20418, U.S.A., February 1973. 90pp.

A basic description of ferrocement and its uses for a wide variety of purposes. This is an excellent introductory handbook and is available free of charge from the above address.

Catchment tanks in Southern Africa — A Review, D.M. Farrar and A.J. Pacey, Africa Field Work and Technology Notes, Oxfam, 274 Banbury Road, Oxford OX2 7DZ, U.K., August 1974. 12pp.

A report on catchment tanks in Southern Africa describing a wide variety of ground and roof catchment tanks. A good basic description of the domestic use of water tanks and an attempt to compare and evaluate the different designs.

Domestic water supplies in rural areas, N. Harris. Geological Survey of Uganda — Water Supply Paper No.3, Geological Survey and Mines Department, P.O.B. 9, Entebbe, Uganda. 1957. 4pp.

The potential of roof catchment tanks for domestic water supplies in the rural areas of Uganda is discussed and costed. Corrugated iron and concrete tanks are described.

The introduction of catchment systems for rural water supplies — a benefit/cost study in a S.E. Ghana village, R.H. Parker, 1972, Dept. of Agricultural Economics and Management, University of Reading, London Road, Reading RG1 5AQ, U.K. 45pp.

An economic analysis of the possibilities of improving village water supplies in Ghana. The analysis gives a useful description of the costs involved in the different approaches of individual or communal water supplies from roof and ground catchments.

The effects of storage on the bacteria of hygienic significance, S.F.B. Poynter and J.K. Stevens, Paper 4, Session 2. A paper presented to the Water Research Centre Conference held at Reading University, U.K., 1975 on water storage and quality. University of Reading, Whiteknights, Reading RG6 2AY, U.K. 22pp.

It describes the ability of stored water to reduce the numbers of pathogenic bacteria present in the water. Although it refers to cold water conditions of less than $20°C$, the results indicate that the storage is an important part of natural water treatment processes and is likely to be even more effective in warmer climates.

www.ingramcontent.com/pod-product-compliance
Lightning Source LLC
Jackson TN
JSHW011405130125
77033JS00023B/854